Tucholsky Wagner Zola Scott Sydow Freud Schlegel
Turgenev Wallace Fonatne
Twain Walther von der Vogelweide Fouqué Friedrich II. von Preußen
Weber Freiligrath Frey
Fechner Ernst Frommel
Fichte Weiße Rose von Fallersleben Kant Richthofen
Hölderlin
Engels Fielding Eichendorff Tacitus Dumas
Fehrs Faber Flaubert
Eliasberg Ebner Eschenbach
Feuerbach Maximilian I. von Habsburg Fock Eliot Zweig
Ewald Vergil
Goethe Elisabeth von Österreich London
Mendelssohn Balzac Shakespeare Ganghofer
Lichtenberg Rathenau Dostojewski
Trackl Stevenson Doyle Gjellerup
Mommsen Tolstoi Hambruch
Thoma Lenz Droste-Hülshoff
Dach Verne von Arnim Hägele Hauff Humboldt
Karrillon Reuter Rousseau Hagen Hauptmann Gautier
Garschin
Damaschke Defoe Hebbel Baudelaire
Descartes
Hegel Kussmaul Herder
Wolfram von Eschenbach Darwin Dickens Schopenhauer Rilke George
Bronner Melville Grimm Jerome Bebel
Campe Horváth Aristoteles Proust
Bismarck Vigny Barlach Voltaire Federer Herodot
Gengenbach Heine
Storm Casanova Tersteegen Grillparzer Georgy
Chamberlain Lessing Langbein Gilm
Brentano Lafontaine Gryphius
Strachwitz Claudius Schiller Kralik Iffland Sokrates
Bellamy Schilling
Katharina II. von Rußland Gerstäcker Raabe Gibbon Tschechow
Löns Hesse Hoffmann Gogol Wilde Vulpius
Luther Heym Hofmannsthal Klee Hölty Morgenstern Gleim
Roth Heyse Klopstock Kleist Goedicke
Luxemburg Puschkin Homer Mörike
La Roche Horaz Musil
Machiavelli Kierkegaard Kraft Kraus
Navarra Aurel Musset Moltke
Nestroy Marie de France Lamprecht Kind Kirchhoff Hugo
Laotse Ipsen Liebknecht
Nietzsche Nansen
Marx Lassalle Gorki Klett Leibniz Ringelnatz
von Ossietzky May Lawrence Irving
Petalozzi vom Stein
Platon Knigge
Sachs Pückler Michelangelo Kock Kafka
Poe Liebermann
de Sade Praetorius Mistral Zetkin Korolenko

The publishing house tredition has created the series **TREDITION CLASSICS**. It contains classical literature works from over two thousand years. Most of these titles have been out of print and off the bookstore shelves for decades.

The book series is intended to preserve the cultural legacy and to promote the timeless works of classical literature. As a reader of a **TREDITION CLASSICS** book, the reader supports the mission to save many of the amazing works of world literature from oblivion.

The symbol of **TREDITION CLASSICS** is Johannes Gutenberg (1400 – 1468), the inventor of movable type printing.

With the series, tredition intends to make thousands of international literature classics available in printed format again – worldwide.

All books are available at book retailers worldwide in paperback and in hardcover. For more information please visit: www.tredition.com

tredition was established in 2006 by Sandra Latusseck and Soenke Schulz. Based in Hamburg, Germany, tredition offers publishing solutions to authors and publishing houses, combined with worldwide distribution of printed and digital book content. tredition is uniquely positioned to enable authors and publishing houses to create books on their own terms and without conventional manufacturing risks.

For more information please visit: www.tredition.com

Object Lessons on the Human Body A Transcript of Lessons Given in the Primary Department of School No. 49, New York City

Sarah F. Buckelew

Imprint

This book is part of the TREDITION CLASSICS series.

Author: Sarah F. Buckelew
Cover design: toepferschumann, Berlin (Germany)

Publisher: tredition GmbH, Hamburg (Germany)
ISBN: 978-3-8491-5084-6

www.tredition.com
www.tredition.de

Copyright:
The content of this book is sourced from the public domain.

The intention of the TREDITION CLASSICS series is to make world literature in the public domain available in printed format. Literary enthusiasts and organizations worldwide have scanned and digitally edited the original texts. tredition has subsequently formatted and redesigned the content into a modern reading layout. Therefore, we cannot guarantee the exact reproduction of the original format of a particular historic edition. Please also note that no modifications have been made to the spelling, therefore it may differ from the orthography used today.

AUTHOR'S NOTE TO THE PUPIL.

This book has been prepared to help you in learning about "the house you live in," and to teach you to take care of it, and keep it from being destroyed by two of its greatest enemies,—Alcohol and Nicotine.

As you study its pages, be sure to find out the meaning of every word in them which you do not understand; for, if you let your tongue say what your mind knows nothing about, you are talking *parrot-fashion*.

And do not forget that you must pay for all the knowledge you obtain, whether you are rich or poor. Nobody else can pay for you. You, your own self, must *pay attention* with your own mind, through your own eyes and ears, *or do without knowledge*.

Be wise: gain all the knowledge you can concerning everything worth knowing, and use it for the good of yourself and other people.

"KNOWLEDGE IS POWER."

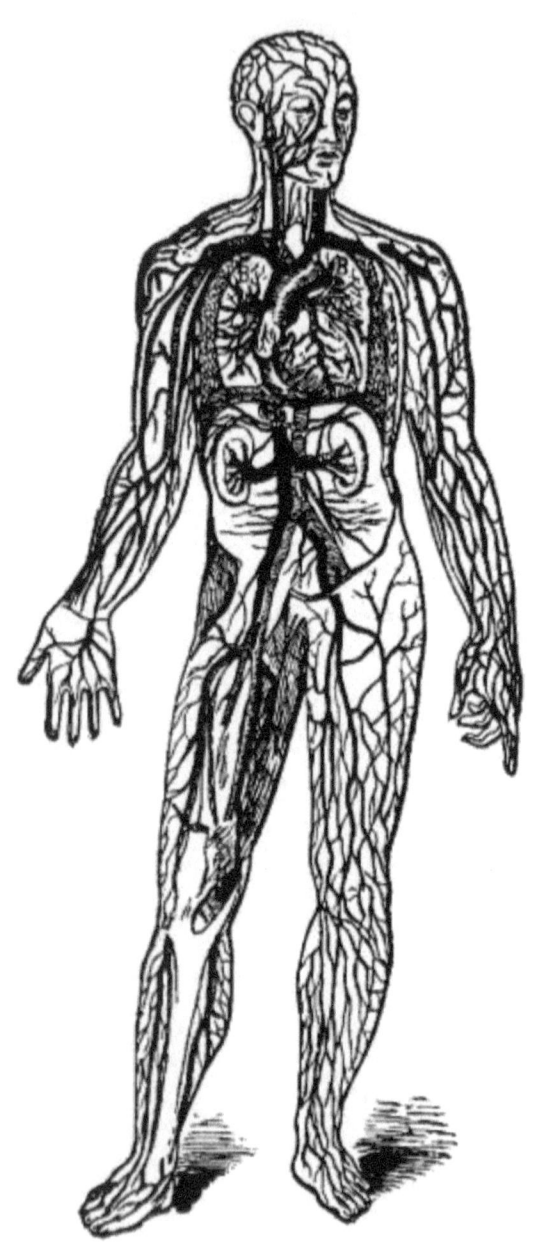

A, the heart; B, the lungs; light cross lines, arteries; heavy lines, veins.

[5]

PART I.

FORMULA FOR INTRODUCTORY LESSONS.

1. My body is built of bones covered with flesh and skin; the blood flows through it, all the time, from my heart. I breathe through my nose and mouth, and take the air into my lungs.

2. The parts of my body are the head, the trunk, the limbs.

3. My head.

The crown of my head.

The back of my head.

The sides of my head.

My face.

My forehead.

My two temples.

My two eyes.

My nose.

My two cheeks.

My mouth.

My chin.

My two ears.

My neck.

My two shoulders.

My two arms.

My two hands.

My trunk.

My back.

My two sides.

My chest.

My two legs.

My two knees.

My two feet.

I am sitting erect.

QUESTIONS FOR THE FORMULA.

1. Tell about your body.

2. Name the parts of the body.

3. Name the parts of the head, trunk, and limbs.

[6]

THE NOSE AND THE MOUTH.

Be sure to keep your mouth closed when you are not talking or singing, especially when you are walking, running, or *asleep*. The two nostrils are outside doors, always open to admit the air, and inside of the upper part of the nose there are two other openings, through which it passes into the throat. Air which goes this way is warmed, cleansed, and moistened, but that which is breathed directly through the mouth is not so well prepared for its work in the lungs.

Do not use your mouth as a box or a pin-cushion; the pin, or whatever yon have put into it, may slip into your throat and cause your death.

QUESTIONS ON THE INTRODUCTORY LESSONS.

Of what is the body built? — "Of bones."

What covers the bones? — "Flesh."

What covers the flesh? — "Skin."

What flows through the body? — "Blood."

Where does the blood flow from? — "The heart."

When does the blood flow from the heart? — "Every time the heart beats."

Show with your hand how the heart beats.

When does the heart beat? — "All the time."

What happens when the heart stops beating? — "We die."

What do you see on the back of your hand, beneath the skin? — "Veins"

What is in the veins? — "Bad blood."

What are the veins? — "Pipes for the bad blood to pass through."

Where do the veins carry the bad blood? — "To the heart."

Where does the heart send the bad blood? — "To the lungs."

What happens to the bad blood when in the lungs? — "It is made pure."

What makes the bad blood pure? — "The air."

How does the air get into the lungs? — "Through my nose, mouth, and windpipe."

[7]

What is breathing? — "Letting the air into and out of my lungs, through my nose, mouth, and windpipe."

When do you breathe? — "All the time."

What do you breathe? — "Air."

What do you breaths through? — "My nose, mouth, and windpipe."

Where do you get the air? — "Everywhere."

Where do the lungs send the pure blood? — "To the heart."

Where does the heart send the pure blood? — "All through the body."

How does the heart send the pure blood through the body? — "Through pipes called arteries."

What kind of blood passes through the arteries? — "Pure blood."

What kind of blood passes through the veins? — "Impure blood."

What carries the pure blood through the body? — "The arteries."

What carries the impure blood through the body? — "The veins."

What makes blood? — "Food and drink."

What is food? — "Anything good to eat."

What is drink? — "Anything good to drink."

Name some kinds of wholesome food. — "Meat, potatoes, oranges, apples, etc."

Name some kinds of wholesome drink. — "Water, milk, lemonade, etc."

What do you mean by wholesome food? — "Food that will make good blood."

What do you mean by wholesome drink? — "Drink that will make good blood."

What does the blood make? — "Bones, flesh, skin, hair, nails, and cartilage."

What use is the blood to the body? — "It makes the body grow, and keeps it alive." [8]

Name some kinds of poisonous drinks. — "Rum, brandy, ale, cider, etc."

What do you mean by poisonous drinks? — "Drinks which hurt or poison the body."

Why do you say that rum and the other drinks you have named are poisonous? — "Because they do harm to every part of the body."

Which part do they hurt most? — "The head or brain."

What harm do they do to the brain? — "They make it unfit to do its work."

What work does the brain do? — "Thinking."

Then what harm do rum, brandy, wine, and these other drinks do to the brain? — "They make it unfit to think."

What other poison do some people use? — "Tobacco."

When do children use tobacco? — "When they chew tobacco; when they smoke cigars or cigarettes."

How much does tobacco poison hurt children? — "More than it hurts anybody else."

In what way does it hurt children? — "It keeps children from growing fast; from being strong and healthy; and from learning as well as they ought."

How does it do all this mischief to children? — "It poisons their lungs, their heart and blood, and their brain."

[9]

PART II.

FORMULA FOR THE PARTS AND JOINTS OF THE BODY:

1. My limbs are my two arms and my two legs.
2. My arm has two parts:

    my upper arm,   my fore-arm;

    and three joints:

    my shoulder joint,   my elbow joint,   my wrist joint.

3. My hand is used in holding, throwing, catching, and feeling:

    the palm of my hand,

    the back of my hand,

    my fingers,

    my thumb,

    my forefinger,

    my middle finger,

    my ring finger,

    my little finger,

    my knuckles,

    my finger joints,

    my nails,

    the tips of my fingers,

    the veins,

    the ball of my thumb,

    and the lines where the flesh is bent.

4. My leg has two parts:

    my thigh,   and my lower leg;

and three joints:

    my hip joint,   my knee joint,   my ankle joint.

5. My foot is used in standing, walking, running, skating, and jumping:

    my instep,

    my toes,

    the sole of my foot,

    the ball,

    the hollow,

    the heel,

    my toe joints,

    and my toe nails, which protect my toes.

[10]

QUESTIONS FOR THE FORMULA.

1. Which are your limbs?
2. Tell about your arm.
3. Tell about your hand.
4. Tell about your leg.
5. Tell about your foot.

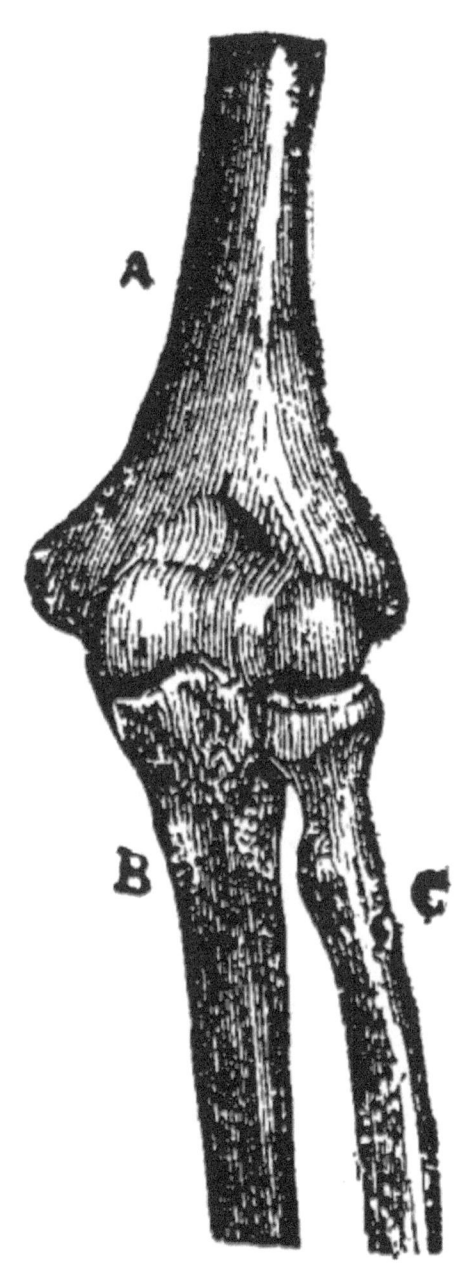

THE EL-

BOW JOINT.
(A hinge joint.)

THE HIP JOINT.
(A ball-and-socket joint.)

Some joints, as those of the skull, are immovable; some, as those of the spine, may be moved a little; and others more or less freely, as those of the limbs. In machines, the parts which move upon each other need to be oiled, to keep them from wearing out; but the joints of our bodies oil themselves with a thin fluid, called *synovia*. This fluid resembles the white of an egg, and comes from a smooth lining inside of the joints. The ends of the bones which form joints are covered by gristle or *cartilage*, and are fastened together by very strong, silvery white bands, called *ligaments*. A sprain is caused by overstretching or tearing some of these ligaments.

[11]

QUESTIONS ON THE LIMBS AND JOINTS OF THE BODY.

What is the trunk of your body? — "All the body but the head and limbs."

Which are your limbs? — "My two arms and my two legs."

How many limbs have you? — "Four."

How many parts has your arm? — "Two parts: my upper arm and my fore-arm."

How many parts has your leg? — "Two parts: my thigh and my lower leg."

How many joints has your arm? — "Three joints: my shoulder joint, my elbow joint, my wrist joint."

How many joints has your leg? — "Three joints: my hip joint, my knee joint, my ankle joint."

What are joints? — "Bending places."

How many kinds of joints have you? — "Two: hinge joints, and ball-and-socket joints."

What kind of a joint is the shoulder joint? — "A ball-and-socket joint."

Why do you call the shoulder joint a ball-and-socket joint? — "Because at the shoulder the arm may move in any direction."

Tell how the shoulder joint is made. — "The upper end of the bone of the upper arm is rounded and fastened in a hollow place called a socket."

Which of the joints of the arm and hand are hinge joints? — "The elbow joint, the wrist joint, the thumb joint, the finger joints."

Which of the joints of the leg and foot are hinge joints? — "The knee joint, the ankle joint, the toe joint."

Which of the joints of the leg is a ball-and-socket joint? — "The hip joint."

Where is the heel? — "At the back part of the foot."

Where is the ball of the foot? — "On the sole of the foot, behind the great toe."

Where is the hollow of the foot? — "In the middle of the sole of the foot." [12]

Where is the sole of the foot? — "On the bottom of the foot."

Where is the instep? — "Between the ankle joint and the toes."

Where is the lower leg? — "Between the knee joint and the ankle joint."

Where is the thigh? — "Between the hip joint and the knee joint."

Where is the upper arm? — "Between the shoulder joint and the elbow joint."

Where is the fore-arm? — "Between the elbow joint and the wrist joint."

Where are the toe joints? — "Between the parts of the toes."

Where are the finger joints? — "Between the parts of the fingers."

Where is the ankle joint? — "Between the lower leg and the foot."

Where is the knee joint? — "Between the thigh and the lower leg."

Where is the hip joint? — "Between the trunk and the thigh."

Where is the wrist joint? — "Between the fore-arm and the hand."

Where is the elbow joint? — "Between the upper arm and the fore-arm."

Where is the shoulder joint? — "Between the trunk and the upper arm."

Where are the tips of the fingers? — "At the ends of the fingers."

Where is the ball of the thumb? — "On the palm of the hand, below the thumb."

Where is the palm of the hand? — "On the inside of the hand, between the wrist and fingers."

[14]

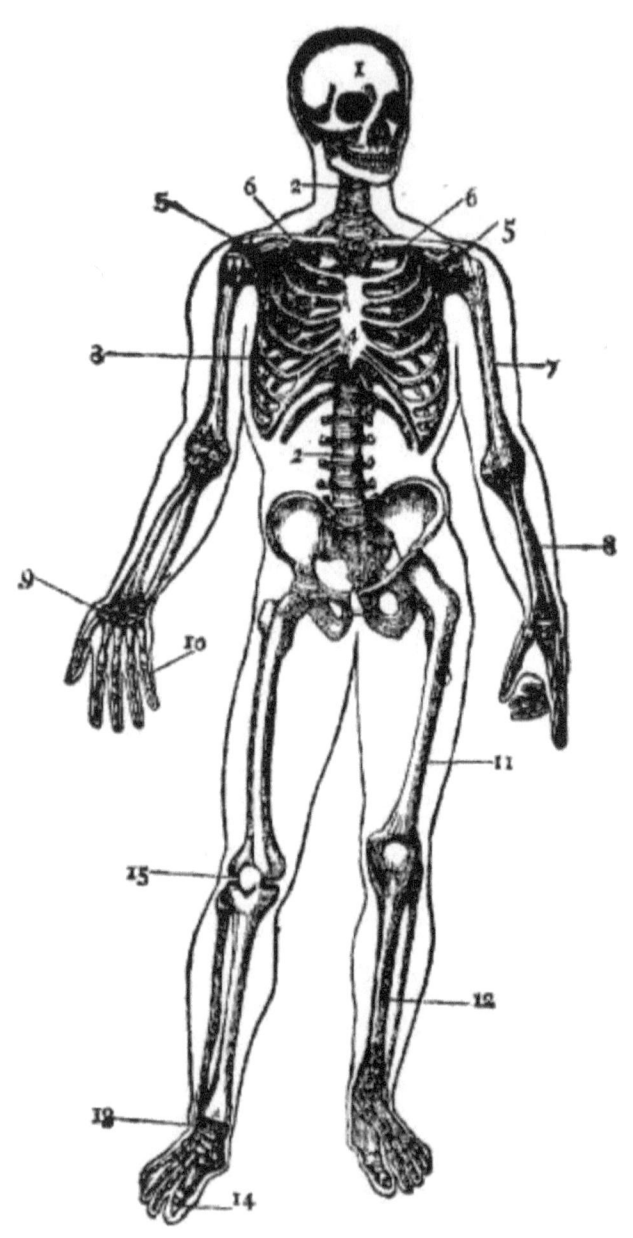

THE SKELETON.
1. The skull.
2. The spine.
3. The ribs.
4. The breastbone.
5. The shoulder blades.
6. The collar bones.
7. The bone of the upper arm.
8. The bones of the forearm.
9. The bones of the wrist.
10. The bones of the fingers.
11. The bones of the thigh.
12. The bones of the lower leg.
13. The bones of the ankle.
14. The bones of the toes.
15. The kneepan.

[15]

PART III.

FORMULA FOR THE LESSON ON THE BONES OF THE BODY.

1. My bones are hard; they make my body strong. There are about two hundred bones in my body.

2. The bones of my head are

>   my skull and my lower jaw;

my face has fourteen bones; my ear has four small bones; at the root of my tongue is one bone.

3. The bones of my trunk are

>   my spine,
>
>   my ribs,
>
>   my breastbone,
>
>   my two shoulder blades,
>
>   and my two collar bones.

4. My upper arm has one bone; my fore-arm has two bones; my wrist has eight bones; from my wrist to my knuckles are five bones; my thumb has two bones; each finger has three bones, making nineteen bones in my hand.

5. My thigh has one bone; my lower leg has two bones; my knee-pan is the cap which covers and protects my knee; in my foot, near my heel, are seven bones; in the middle of my foot are five bones; my great toe has two bones; each of my other toes has three bones; making twenty-six bones in my foot.

QUESTIONS FOR THE FORMULA.

1. Tell about your bones.

2. Tell about the bones of the head.

3. Tell about the bones of the trunk. [16]

4. Tell about the bones of the arm and hand, beginning with the upper arm.

5. Count the bones of the hand.

6. Tell about the bones of the leg and foot, beginning with the thigh.

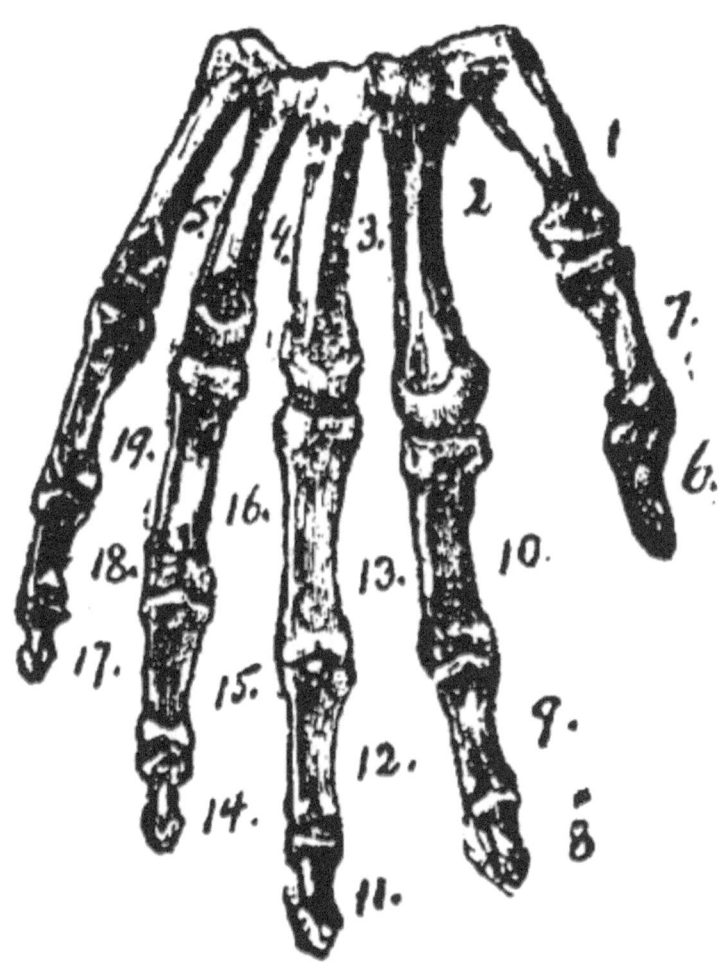

FIG. B.

    1, 2, 3, 4, 5, the bones of the palm of the hand.

    6, 7, the bones of the thumb.

    8, 9, 10, the bones of the first or fore-finger.

    11, 12, 13, the bones of the second or middle finger.

14, 15, 16, the bones of the third or ring finger.
17, 18, 19, the bones of the fourth or little finger.

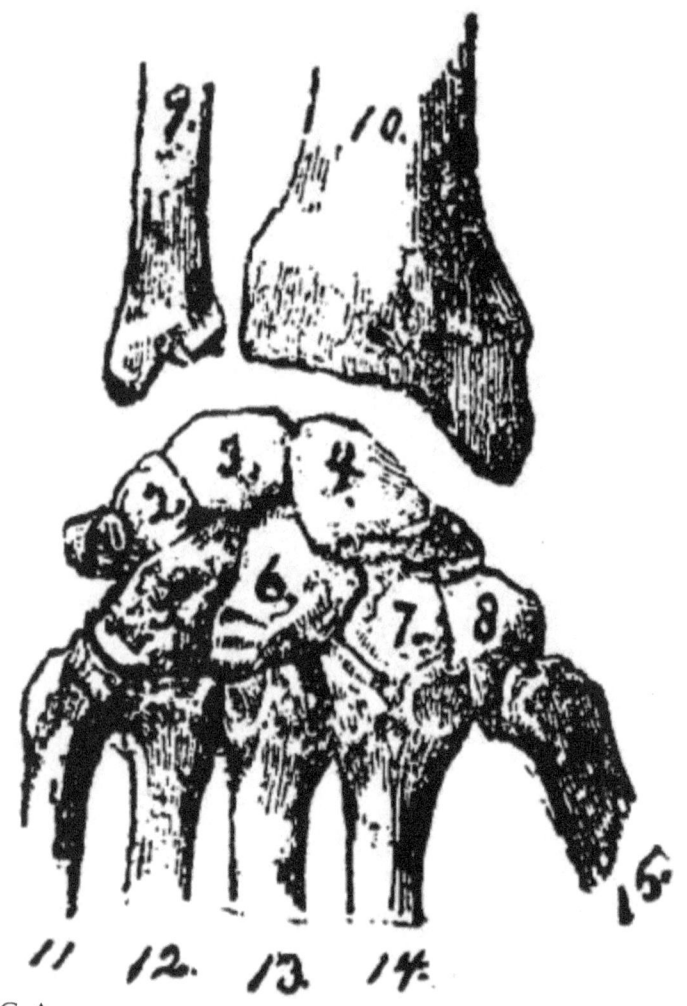

FIG. A.

1, 2, 3, 4, the upper row of the bones of the wrist.
5, 6, 7, 8, the lower row of the bones of the wrist.

9, 10, the lower ends of the bones of the fore-arm.

11, 12, 13, 14, 15, the upper ends of the bones of the palm of the hand.

The bones of the wrist are so firmly fastened together that they are seldom put out of place. The upper row joins with the bones of the fore-arm, the lower with those of the palm of the hand.

[17]
QUESTIONS ON THE BONES.

How many bones in the body? — "About two hundred."

Of what use are the bones to the body? — "They make the body strong; they form the framework of the body."

How many bones in the face? — "Fourteen."

How many bones in the ear? — "Four small bones."

How many bones at the root of the tongue? — "One."

How many bones in the upper arm? — "One."

How many bones in the fore-arm? — "Two."

How many bones between the wrist and the knuckles? — "Five."

How many bones in the thumb? — "Two."

How many bones in each of the fingers? — "Three."

How many bones in the whole hand? — "Nineteen."

How many bones in the hand and arm? — "Thirty."

How many bones in the thigh? — "One long bone."

How many bones in the lower leg? — "Two."

How many bones in the heel? — "Seven."

How many bones in the middle of the foot? — "Five."

How many bones in the great toe? — "Two."

How many bones in each of the other toes? — "Three."

How many bones in the whole foot? — "Twenty-six."

How many bones in the foot and leg? — "Thirty."

How many bones in two arms and two hands? — "Sixty."

How many bones in two legs and two feet? — "Sixty."

How many bones in the limbs? — "One hundred and twenty."

Where is the knee-pan? — "Over the knee joint."

Where is the longest bone of the body? — "In the thigh."

Where are the smallest bones of the body? — "In the ear."

Point to the collar bones.

Point to the shoulder blades.

How many collar bones have you? — "Two."

How many shoulder blades have you? — "Two."

Point to the spine.

Point to the breastbone.

Point to the skull.

[18]
EXERCISE FOR COUNTING THE BONES OF THE HAND.
FOR PRIMARY CLASSES.
I.

1. Close both hands.

2. Raise the forefinger of the right hand, as the index or pointing finger.

3. Place the index finger upon the lower thumb joint of the left hand.

4. Draw the index finger down to the wrist, over the bone between the thumb knuckle and the wrist, and count "One."

5. Place the index finger on the knuckle of the first finger.

6. Draw the index finger down to the wrist, over the bone leading from the first finger to the wrist, and count "Two."

7. So on, for each of the three other bones of the hand. Repeat until no mistake is made in touching or counting.

II.

1. Raise the thumb, and place the index finger of the right hand on the middle of the upper part of the thumb for bone "Six"; then

2. On the lower part of the thumb for bone "Seven." Repeat from the beginning, until the children can touch and count each bone properly.

III.

1. Keep the thumb erect; raise the first finger of the left hand.

2. Place the index finger on the bone between the tip and the first joint of the first finger for bone "Eight."

3. Between the first and middle joint for bone "Nine."

4. Between the middle and third joint for bone "Ten." Review, from the beginning, until the class can touch and count every bone as directed.

IV.

1. Keep the thumb and forefinger erect; raise the second finger and touch, as in the lesson on the first finger bones, "Eleven," "Twelve," and "Thirteen." Review. [19]

2. Proceed in the same manner for the third and fourth fingers, always beginning with the bone nearest the tip of the finger, and touching that at the lowest part last.

If the exercise has been properly performed, every child will say "Nineteen" as its index finger touches the lowest bone of the little finger, and all the fingers of every left hand will be outspread.

THE BONES

OF THE HEAD:
Skull	8
Face, including the lower jaw	14
Tongue	1
Ears	8
	——
	31

OF THE TRUNK:
Spine	24
Ribs	24
Breastbone	8
Shoulder blades	2
Collar bones	2
	——
	60

OF THE UPPER LIMBS:
Upper arms	1 x 2 = 2
Fore-arms	2 x 2 = 4
Wrists	8 x 2 = 16
Hands	19 x 2 = 38
	——
	60

OF THE LOWER LIMBS:
Thighs	1 x 2 = 2
Knee-pans	1 x 2 = 2
Lower legs	2 x 2 = 4
Feet	26 x 2 = 52
	——
	60

Total, 211, not including the teeth.[1]

We teach the children to say "about two hundred," because there is not always the same number of bones in the body. In some parts two or three bones unite and form one bone. For example: the breastbone of a child is made up of eight pieces; some of these unite as it becomes older, so that when fully grown it has but three pieces in this bone.

[1] The teeth are not bone, but a kind of soft, bone-like substance, called *dentine*. Common ivory is dentine.

PART IV.

FORMULAS FOR THE LESSONS ON THE ORGANS OF SENSE.

1. *The Eyes.* — My eyes are to see with.

My eye is like a ball in a deep, bony socket. The black circle in the centre is the pupil or window of my eye; the colored ring is the iris or curtain; the white part is the eyeball.

My upper and lower eyelids cover and protect my eyes.

My eyebrows are for beauty, and keep the perspiration from rolling into my eyes.

My eyes are washed by teardrops every time I wink my eyelids.

2. *The Ears.* — My ears are to hear with:

> the rim of my ear,
>
> the flap of my ear,
>
> the drum of my ear.

The drum of my ear is protected by a fence of short, stiff hairs, and by a bitter wax about the roots of these hairs.

3. *The Nose.* — My nose is to smell and breathe with; it is in the middle of my face:

> my two nostrils,
>
> the bridge of my nose,
>
> the cartilage,
>
> the tip of my nose.

My nostrils lead to a passage back of my mouth through which I breathe.

The cartilage separates my nose into two parts.

4. *The Mouth.* — My mouth is to speak, eat, and breathe through:

my upper lip,

my lower lip.

[21]

In my mouth are:

my tongue,

my lower teeth,

my upper teeth,

my lower teeth,

and my upper and lower jaws, covered with flesh called *gum*.

5. *The Teeth.* — My teeth are used in eating and talking.

My teeth are made of a soft kind of bone, covered with enamel.

I have three kinds of teeth: cutting teeth, tearing teeth, grinding teeth.

A young child has twenty teeth, ten in each jaw.

A grown person has thirty-two teeth, sixteen in each jaw.

6. To preserve my teeth:

I must keep them clean.

I must not scratch the enamel.

I must not eat or drink anything very hot or very cold.

I must not use them for scissors or nut-crackers.

I must not burn them with tobacco or cigars.

7. *About Eating.* — When I eat I move my lower jaw only.

My tongue brings the food between my teeth,

the cutters cut it,

the tearers tear it,

the grinders grind it,

the saliva moistens it,

and my tongue helps me to swallow it.

QUESTIONS FOR THE FORMULAS.

1. Tell about your eyes.
2. Tell about your ears.
3. Tell about your nose.
4. Tell about your mouth.
5. Tell about your teeth.
6. What is necessary if you would preserve your teeth?
7. Tell about eating.

[22]

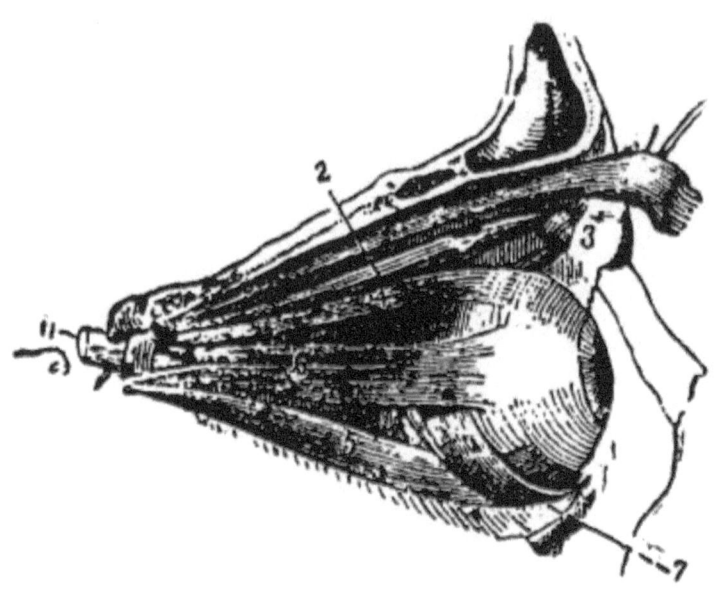

1, the muscle which raises the upper eyelid.

2, the upper oblique muscle.

7, the lower oblique muscle. The oblique muscles roll the eye inward and downward.

4, 5, 6, three of the *four* straight muscles. Two of the straight muscles roll the eye up and down; the other two move it right and left.

3, the pulley through which the upper oblique muscle plays.]

QUESTIONS ON THE DESCRIPTION OF THE EYES.

Of what shape is the eye? — "It is round like a ball."

In what is it placed? — "In a deep, bony socket."

What is a socket? — "A hollow place."

Why is the eye placed in a deep, bony socket? — "To keep it from getting hurt."

Why would not an eye shaped like a cube do for us? — "It would not look well; it could not be rolled about."

Why would not an eye shaped like a cone or cylinder do for us? — "It could not be rolled in every direction."

Why is the ball-shape best for the eye? — "It looks best, and may be rolled in every direction."

What part of the eye do we see through? — "The black spot in the centre."

What is it called? — "The pupil."

[23]

What shape is the pupil? — "Round like a circle."

What color is the pupil? — "Black."

Of what use is the pupil? — "To let light into the eye; to see through."

What is around the pupil? — "A colored ring."

What is the colored ring called? — "The iris."

Of what use is the iris? — "It acts like a curtain to the eye; it lets in and keeps out light from the pupil."

Of what shape is the iris? — "Round like a ring."

Of what color is the iris? — "Sometimes blue, sometimes brown, sometimes gray."

Does the iris always appear the same in size? — "It does not: sometimes it looks large, sometimes small."

When is it the largest? — "When it rolls over the pupil to keep out the strong light."

When is it the smallest? — "When it rolls backward, to let light into the pupil."

When is the pupil the largest? — "When we are in the dark."

When is the pupil the smallest? — "When we are in a bright light."

What color is the eyeball? — "White."

What shape is the eyeball? — "Round like a ball."

How is the eyeball held in its socket? — "By cords made of flesh."

Where are the eyebrows? — "Above the eyelids."

Of what use are the eyebrows? — "To keep the perspiration from rolling into the eyes."

Where are the eyelids? — "Over the eyes."

Of what use are they? — "They cover the eyes and keep them from getting hurt."

Where are the eyelashes? — "On the edges of the eyelids."

Of what use are the tears? — "They keep the eyes clean; they make the eyes move easily in their sockets."

Where are the tears made? — "Back of the eyebrows."

When do the tears wash the eyes? — "Every time we wink our eyelids."

[24]
QUESTIONS ON THE EARS.

Name the parts of the ear.

Where are your ears? — "On the sides of my head."

Which is the rim of the ear? — "The edge of the ear."

Which is the flap of the ear? — "The lower part of the ear."

Where is the drum of the ear? — "Inside of the ear."

How is the drum protected? — "By stiff hairs and a bitter wax at its entrance."

QUESTIONS ON THE NOSE.

Where is the nose? — "In the middle of the face."

Name the parts of the nose.

Where is the tip of the nose? — "At the end of the nose."

Where is the bridge of the nose? — "At the top of the nose, between the eyes."

Where is the cartilage? — "In the middle of the inside of the nose."

Of what use is the nose? — "To smell and breathe through."

What are the nostrils? — "The openings inside of the nose."

Of what use are the nostrils? — "To let the air into and out of the opening back of the mouth."

QUESTIONS ON THE MOUTH, ETC.

Where is the mouth? — "In the lower part of the face, between the nose and the chin."

Of what use is the mouth? — "To breathe, speak, and eat through."

What is in the mouth? — "My tongue, my upper teeth, my lower teeth, and my upper and lower jaws."

What covers the jaws? — "Red flesh, called *gum*."

Of what are the jaws composed? — "Of bones."

Of what are the teeth made? — "Of dentine, covered with enamel." See note, p. 19.

[25]

What is enamel? — "A smooth, white substance, harder than bone."

Of what use are the teeth? — "To eat and talk with."

What kinds of teeth have you? — "Cutting teeth, tearing teeth, grinding teeth."

Describe the cutting teeth. — "The cutting teeth have broad and flat edges."

Describe the tearing teeth. — "The tearing teeth are sharp and pointed."

Describe the grinding teeth. — "The grinding teeth are the thick, back teeth."

Which jaw is moved in eating? — "The lower jaw."

What work do the teeth perform? — "They cut, tear, and grind the food."

How many teeth has a child in a full set? — "Twenty teeth: ten in each jaw."

How many teeth has a grown person in a full set? — "Thirty-two: sixteen in each jaw."

What does the tongue do in eating? — "It rolls the food between the teeth, and helps in swallowing."

What is the saliva? — "A kind of liquid, sometimes called *spit*."

Of what use is it in eating? — "It wets and softens the food."

What do you mean by preserve? — "To keep from injury."

What do you mean by injury? — "Hurt."

How do you preserve your teeth? See Formula.

How do very hot or very cold drinks hurt the teeth? — "They crack the enamel."

What happens if the enamel is cracked? — "The teeth decay."

Then what must you do to preserve your teeth? — "I must try to keep the enamel from being cracked or injured in any way."

[26]

PART V.

FORMULA FOR DESCRIPTION OF THE BONES.

1. My skull is formed of several bones united, like two saws with their toothed edges hooked into each other.

2. My spine extends from the base of the skull behind, down the middle of my back.

It is composed of twenty-four short bones, piled one upon the other, with cartilage between them.

These bones are fastened together, forming an upright and flexible column, which makes me erect and graceful.

3. My ribs are curved, strong, and light; there are twenty-four of them, twelve on each side; they are fastened at the back to my spine, in front to my breastbone, forming a hollow place for my heart, lungs, and stomach.

4. My shoulder blades are flat, thin, and like a triangle in shape; they are for my arms to rest upon.

5. My collar bones are fastened to my shoulder blades and my breastbone; they keep my arms from sliding too far forward.

6. The bones of old people are hard and brittle; those of children soft and flexible; so I must sit and stand erect, that mine may not be bent out of shape. I must not wear tight clothing, or do anything that will crowd them out of their places.

7. My bones are made from my food, after it has been changed into blood; so I must be careful to eat good, wholesome food, that they may be strong and healthy.

8. I must not breathe impure air, because impure air makes bad blood, and bad blood makes poor bones.

9. The body of every person is changing all the time, because the skin, flesh, and bones are always wearing out, and the blood is always repairing and building them again.

QUESTIONS FOR THE FORMULA.

1. Tell about the skull.

2. Tell about the spine.

3. Tell about the ribs.

4. Tell about the shoulder blades.

5. Tell about the collar bones.

6. Tell about the difference between the bones of old people and those of children.

7. Of what are your bones made?

8. If you wish your bones to be strong, why should you not breathe impure air?

9. What have you learned about the change which is always taking place in the body?

THE JOINTS OF THE SKULL.

[28]

A little girl was looking at some pictures of ladies in fashionable dresses. While admiring the beautiful styles and bright colors of the garments, she pointed to the waist of one, and exclaimed, "*That means trouble.*" The waist was too small for a grown person, and could only have been made so by *tight-lacing*. The child had been taught that dresses, corsets, coats, vests, bands, or anything fastened tightly around the waist, press upon the ribs and crowd them out of place, preventing the heart, lungs, and other inside organs from working as they should, causing headache, dyspepsia, shortness of breath, and often ending in some incurable disease, so she knew that *tight clothing means trouble* to the wearer.

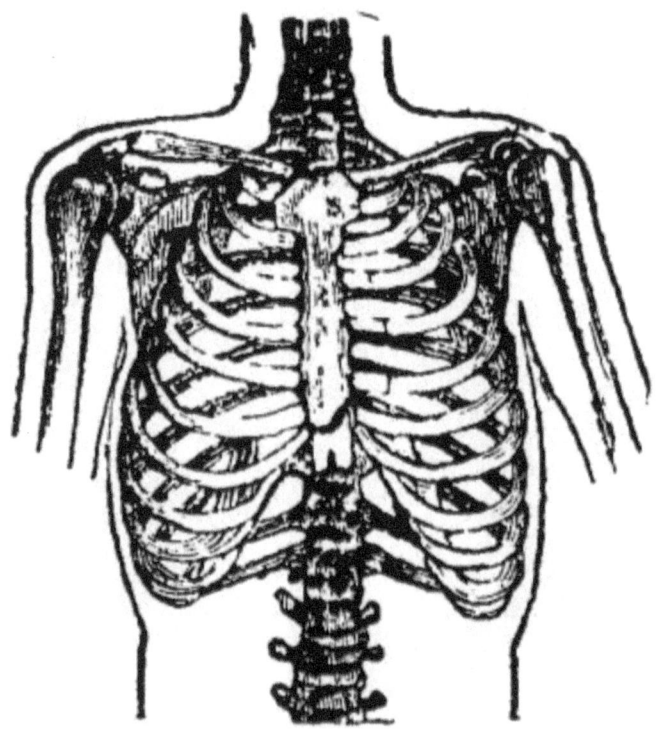

FIG. 2. A natural, well-shaped chest.

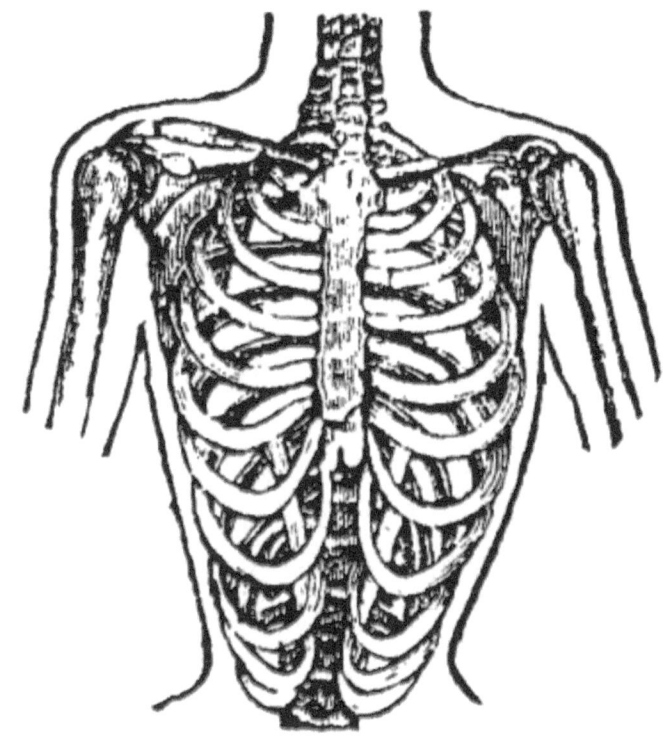

FIG. 1. Deformed by tight-lacing.

QUESTIONS ON THE DESCRIPTION OF THE BONES.

Point to the skull.

Of what is it made? — "Several bones united together."

How are the skull bones united? — "Like two saws with their toothed edges hooked into each other."

What do you mean by *toothed*? — "Having points, like teeth."

What covers the skull? — "Flesh, skin, and hair."

Of what use is the skull? — "It protects the brain."

What is the brain? — "That part of my body in which the thinking is done."

[29]

Where is the spine? — "It extends from the base of my skull behind, down the middle of my back."

What do you mean by *extends*? — "Goes from."

What do you mean by *base*? — "The lower part of anything."

Of what is the spine made? — "Of about twenty-four short bones, with cartilage between them."

What is cartilage? — "An elastic substance, harder than flesh, but softer than bone."

How are the bones of the spine placed? — "They are piled one upon the other."

What do you mean by *forming*? — "Making."

What do you mean by *upright*? — "In a vertical position."

What do you mean by *flexible*? — "Easily bent."

What do you mean by *column*? — "A pillar."

What do you mean by *erect*? — "In a vertical position."

Why is cartilage placed between the bones of the spine? — "To make the spine flexible; to keep the brain from injury when we walk or run."

What do you mean by *elastic*? — "Springing back after having been stretched, squeezed, twisted, or bent."

Tell about your ribs. — "My ribs are curved, strong, and light."

Where are your ribs? — "On each side of my trunk."

How many ribs have you? — "Twenty-four; twelve on each side."

How are your ribs fastened? — "At the back to my spine; in front to my breastbone."

What do your ribs form? — "A hollow place for my heart, lungs, and stomach."

Where are your shoulder blades? — "In the upper part of my back."

What shape are they? — "Flat, thin, and like a triangle."

Of what use are your shoulder blades? — "For my arms to rest upon."

Point to your collar bones.

Where are they fastened? — "To my shoulder blades and my breastbone."

[30]

Of what use are your collar bones? — "They keep my arms from sliding too far forward."

Of what are your bones made? — "Of food after it has been changed into blood."

Why should you eat wholesome food? — "That my bones may be strong and healthy."

How does impure air hurt the bones? — "Impure air makes bad blood, and bad blood makes poor bones."

Why should you sit and stand erect? — "Because my bones are easily bent out of shape; if I do not sit and stand erect, they will grow crooked."

Why is it wrong to wear tight clothing? — "Because tight clothing crowds the bones out of shape."

Whose bones are the more brittle, those of a child, or those of an old person? — "Those of an old person."

What do you mean by *brittle*? — "Easily broken."

Whose are the more flexible? — "Those of a child."

What do you mean by *flexible*? — "Easily bent."

What repairs the worn out bones, flesh, and skin of the body? — "The blood."

What do you mean by *repairs*? — "Mends."

What causes the bones, flesh, and skin of your body to change often? — "The bones, flesh, and skin are always wearing out, and the blood is always building and repairing them again."

What are alcoholic liquors? — "Liquors which have alcohol in them."

Name some alcoholic liquors. — "Beer, wine, rum, etc."

Whose bones mend the more easily when broken, the bones of those who drink alcoholic liquors, or those of the people who do not use these poisons? — "The bones of those who *do not* use alcoholic liquors."

What other poison hurts the bones? — "Tobacco."

How do alcohol and tobacco hurt the bones? — "They make bad blood, and bad blood makes poor bones."

[32]

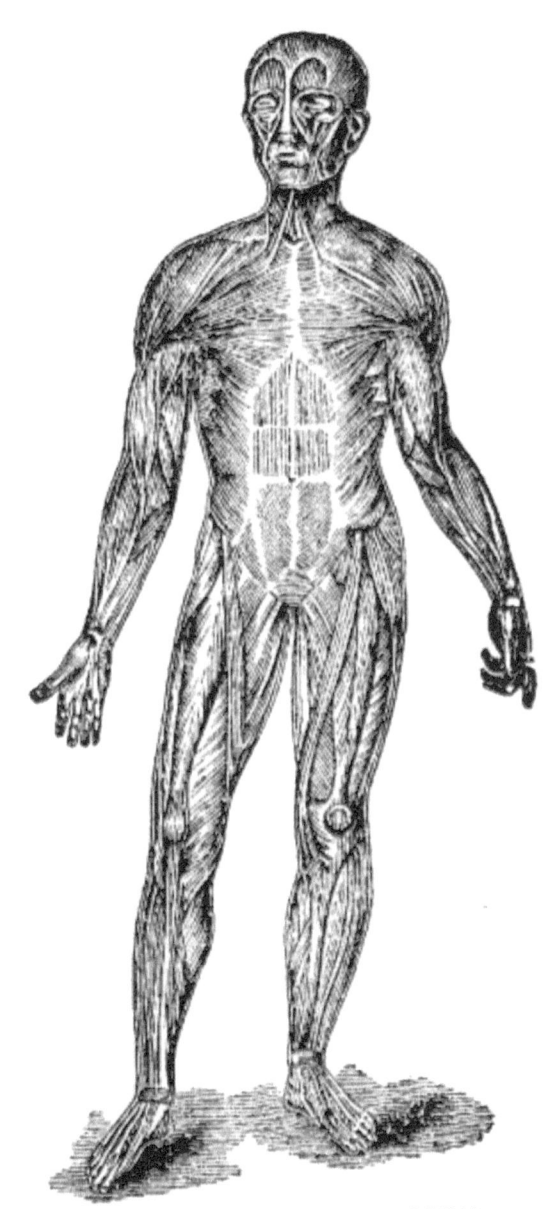

FRONT VIEW OF THE MUSCLES OF THE BODY.

[33]

PART VI.

FORMULA FOR THE LESSON ON THE MUSCLES.

1. Muscles are the red, elastic bands and bundles of thread like substance, called flesh, which cover the bones and make the eyeballs, the eyelids, the tongue, the heart, the lungs, and various other parts of the body.

2. There are about four hundred and fifty muscles in my body.

3. The work of the muscles is to support and move my bones, and different parts of the body.

4. The muscles may be named the muscles of my head, the muscles of my trunk, the muscles of my limbs.

5. The muscles of my head cover and move the parts of my head and face. The muscles of my trunk cover and move the parts of my neck and trunk. The muscles of my limbs cover and mote the parts of my arms and legs.

6. Those muscles are the weakest which I use least; those muscles are the strongest which I exercise most in work or play.

7. If I would be strong and healthy,

my muscles must be used,

my muscles must be rested,

my muscles must be supplied with good blood.

I must exercise in work and play to make them strong; I must sleep, or change my kind of work or play, to give them rest, when they are tired; I must breathe pure air, take wholesome food and drink, and live in the sunlight, to supply them [34] with good blood; I must not weaken them by using alcohol or tobacco.

QUESTIONS FOR THE FORMULA.

1. Tell about the muscles.

2. How many muscles have you in your body?

3. Of what use are the muscles?

4. How may the muscles be named?

5. Tell about the muscles of the head, trunk, and limbs.

6. Which muscles are the weakest, and which are the strongest?

7. What is necessary if you would have strong and healthy muscles?

CLASSES AND WORK OF THE MUSCLES.

The muscles are divided into two great classes: those which we may move as we choose, called *voluntary* muscles, and those over which we have no power, called *involuntary* muscles.

Some muscles support and move the various parts of the body, others have different work to do. The heart, the great involuntary muscle, acts like an engine to drive the blood throughout the body; the lungs draw in and throw out the air in breathing; the stomach helps to churn and change food into blood; the tongue is used in speaking and eating.

QUESTIONS ON THE MUSCLES.

What are the muscles? — "The lean flesh of the body; bands and bundles of fleshy threads which cover the body."

Of what use are the muscles to the body? — "They cover the bones; they support and move the bones and different parts of the body."

Name some parts of the body which are made of muscles. — "The eyeballs, the eyelids, the tongue, the heart, the lungs."

What color are the muscles? — "Red."

How do the muscles move the bones? — "By shortening or lengthening themselves according to the way the bones are to be moved."

Tell how the muscles move your arm at the elbow. — "The [35] muscles in the front part of the arm shorten themselves, to draw my fore-arm toward the shoulder; when I wish to stretch out the fore-arm these muscles lengthen, while another set of muscles shorten, to draw the fore-arm away from the upper arm."

What do you say about the muscles because they have the power to shorten and lengthen themselves? — "They are elastic."

About how many muscles are there in your whole body? — "About four hundred and fifty."

How may these be divided as you study about them? — "They may be divided into the muscles of my head, the muscles of my trunk, and the muscles of my limbs."

Of what use are the muscles of your head? — "They cover and move the parts of my head and face."

Of what use are the muscles of your trunk? — "They move the parts of my neck and trunk."

Of what use are the muscles of your limbs? — "They move the parts of my arms and legs."

How can you make your muscles strong? — "By using them."

How can you make your muscles weak? — "By not using them."

What is necessary to make your muscles strong and healthy? — "They must be used; they must be rested when tired; they must be supplied with pure blood."

How should the muscles be used? — "They should be exercised in work or play."

How may they be rested? — "I may rest my muscles by changing position; by changing my kind of work or play; or by going to sleep."

Explain what you mean by changing your position. — "If I am standing, I must sit or lie down to rest them; if they are tired, because I have been sitting too long, I must rest them by standing, walking, or running."

What do you mean by changing the kind of work or play? — "If, in my work or play, my arms become tired, I must do something in which my arms may rest, though other parts of my body may be in exercise."

[36]

How may you help supply your muscles with good blood? — "By breathing pure air; by taking wholesome food and drink; and by living in the sunlight."

How does drinking alcoholic liquors hurt the muscles? — "It makes them weak, and unfit to move the parts of the body."

What wonderful muscle moves without your will? — "The heart."

How does alcohol hurt the heart? — "It makes it beat too fast."

How does "beating too fast" hurt the heart? — "It makes it tired, and sometimes wears it out." See Appendices on Alcohol and Tobacco.

[38]

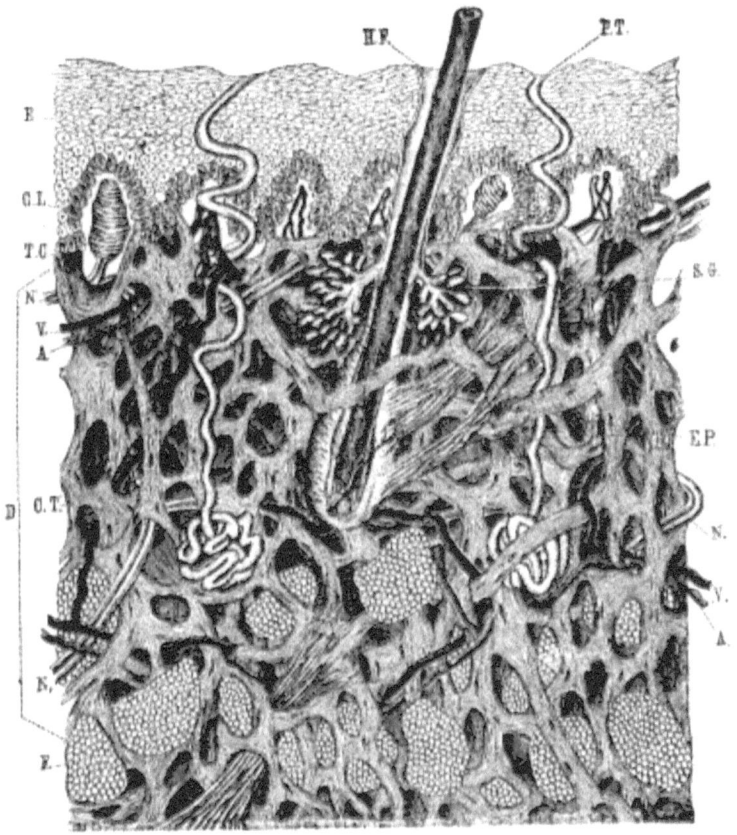

THE SKIN (very highly magnified).—(From Walker's *Physiology*, 1884.)

A, arteries; V, veins; N, nerves; F, fat cells; E, the outer skin; CL, the color layer; D, the true skin; PT, a perspiratory tube; HF, a hair and hair sac; EP, muscles; SG, oil glands; TC, tactile corpuscles; CT, connective tissue.

[39]

PART VII.

FORMULA FOR THE LESSON ON THE SKIN.

1. My skin covers my body.

2. It is thin, elastic, flexible, porous, and absorbent.

3. I have two skins; the inner skin is the true skin.

4. My true skin is elastic, and like a net-work of blood-vessels and nerves. My true skin is covered with a jelly-like substance which gives color to my skin.

5. My outside skin is not the same thickness over my whole body. In some parts, as on the palms of my hands and the soles of my feet, it is very thick and tough.

6. If my outside skin be destroyed, it will grow again; if the jelly-like substance be destroyed, it will re-appear; but if my true skin be destroyed, it will never be perfectly renewed.

7. More than half of the waste substance of my body passes from it through the pores of the skin, in the form of perspiration.

8. If I would have a healthy skin,

I must perspire freely all the time,

I must keep my body clean,

I must wear clean clothing,

I must breathe pure air,

and live in the sunlight.

QUESTIONS FOR THE FORMULA.

1. Where is your skin?

2. Tell about the skin.

[40]

3. How many skins have you?

4. Tell about the true skin.

5. What difference is there in the thickness of your outside skin?

6. What happens if the different skins be destroyed?

7. What passes through the pores of the skin?

8. What is necessary if you would have a healthy skin?

DIRECTIONS FOR BATHING.

Bathe the whole body at least twice every week. Do not bathe when tired or after a hearty meal. After bathing *rub well* with a coarse towel.

QUESTIONS ON THE SKIN.

Of what use is the skin? — "It covers the muscles of the body."

What can you tell about it? — "It is flexible, elastic, porous, and absorbent."

Why do you say it is flexible? — "Because it is easily bent."

Why do you say it is porous? — "Because it is full of little holes, or pores."

Why do you say it is elastic? — "Because it will spring back after it is stretched, squeezed, twisted, or bent."

Why do you say it is absorbent? — "Because it will soak up liquids."

How many skins have you? — "Two; an outside skin, and an inner skin."

Which is the true skin? — "The inner skin."

Of what is the inner skin composed? — "Of blood-vessels and nerves."

How do you know that the outer skin has no blood-vessels? — "Because if I put a pin through the outer skin the blood does not flow out, as it would if I had cut a blood-vessel."

How do you know the outer skin has no nerves? — "Because [41] if I put a pin through my outer skin it does not make me suffer pain, as it would if I had touched a nerve."

What gives color to the skin? — "A jelly-like substance between the inner and the outer skin."

What have you learned about the true skin? — "That it is of the same color in people of every nation."

What difference is there in the thickness of the outer skin? [See Formula.]

What passes through the pores of the skin? [See Formula.]

What is this waste called when it comes from the surface of the skin? — "Perspiration."

When does the perspiration flow through the pores of the skin? — "All the time, if the skin is healthy."

Why do we not always see the perspiration which passes through the pores? — "Because it does not always form drops on the surface of the skin; it generally passes off in very fine particles."

What becomes of the fine or minute portions of perspiration which pass from the body? — "Some of these portions are absorbed by the clothing; some pass into and mix with the air around us."

What effect does the perspiration produce on the air and the clothing? — "It soon makes the air unfit to be breathed, and the clothing unfit to be worn."

What is necessary if you would have a healthy skin? [See Formula.]

Why must you wear clean clothing? — "That there may be nothing impure in the clothing for the pores of the skin to absorb."

Why should you breathe pure air? — "Because air purifies the blood, and pure blood is necessary to make a healthy skin."

How does drinking alcoholic liquors hurt the skin?—"It makes the blood impure, and impure blood makes unhealthy skin."

In what other way does drinking these liquors hurt the skin?—"It gives the skin too much work to do."

[42]

How does it give it too much work to do?—"It makes more waste substance to pass from it through the pores, in the form of perspiration."

In what other way does drinking alcoholic liquors hurt the skin?—"It makes it a bad color."

How does it make the skin a bad color?—"It stretches the little blood-vessels of the skin, and makes them too full of blood." See Appendix.

[44]

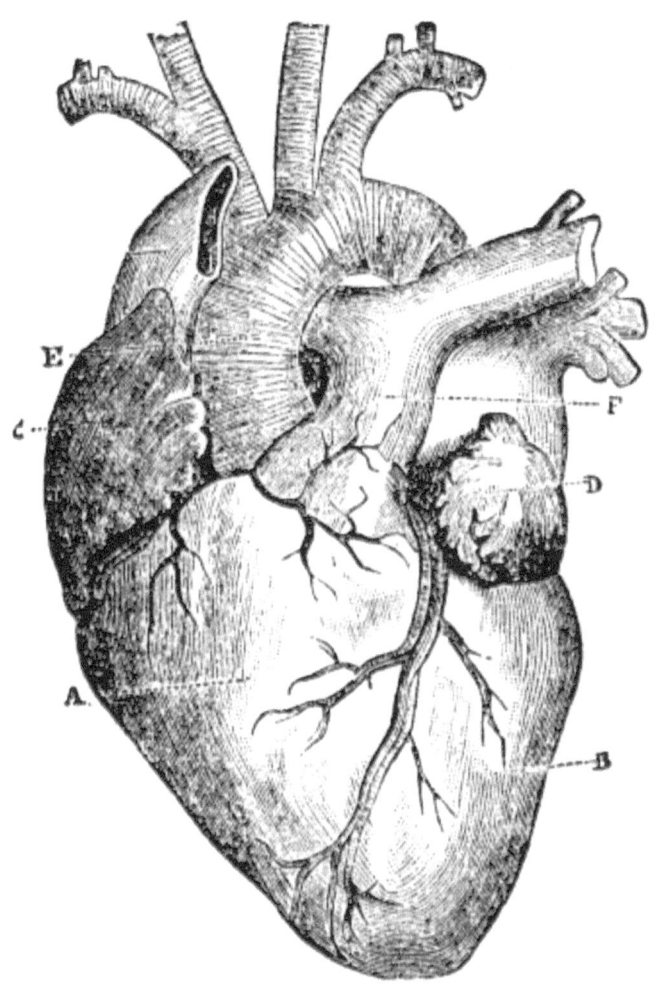

THE HEART.

A, the right ventricle; B, the left ventricle; C, the right auricle D, the left auricle; E, the aorta; F, the pulmonary artery.

[45]

PART VIII.

FORMULA FOR THE LESSON ON THE HEART AND THE CIRCULATION OF THE BLOOD.

1. My heart is shaped like a cone, and placed in my chest near my breastbone, with its apex pointing downward to my left side. It beats about seventy times a minute, sending out about two ounces of blood at every beat.

2. The blood when pure is of a bright red color; it is a liquid made from food and drink.

3. It passes from my heart to all parts of my body, through pipes called arteries; these arteries spread out through the body like branches from a tree.

4. As the blood flows from the heart, through the arteries, it gives nourishment to every part of the body, and carries away the impurities it meets, which makes it black and thick; when it comes through the veins, back to the heart, it is not fit to be used, so it goes to the lungs to be purified by the fresh air; then it returns to the heart to be sent again throughout the body; this happens once in from three to eight minutes, and is called the circulation of the blood.

7. If I would be healthy,

my blood must be pure and circulate freely all the time.

8. It will not circulate freely,

if I wear tight clothing,

if I do not exercise in work or play,

if I do not keep my body warm.

9. It will be impure,

if I breathe bad air,

if I eat unwholesome food,

if I drink alcoholic liquors,

if I snuff, smoke, or chew tobacco.

QUESTIONS FOR THE FORMULA.

1. Tell about the heart and where it is placed.

2. Tell about the blood and of what it is made.

3. Where does the good blood pass after it is sent out from the heart?

4. Tell what the blood does as it flows through the body.

5. What is this flowing of the blood to and from the heart called?

6. How often does it happen?

7. What is necessary if you would have pure blood?

8. When will the blood not circulate freely?

9. When will the blood be impure?

HOW TO TREAT A WOUND.

If it is only a flesh-wound or slight cut, wash it with cold water and bandage it with a clean, white rag. The edges of a deep cut should be drawn together and held in place by narrow strips of adhesive plaster, fastened across the wound from side to side.

If the cut is very deep, and the blood flows very freely, send for a doctor. While you wait for him, knot a handkerchief, or suspender, or towel, in the middle, and twist it very tightly *over the cut artery, above the wound*. If a vein has been severed, twist the knotted handkerchief *below the wound*. If the blood continues to flow, tie a bandage both above and below the hurt part.

[47]

QUESTIONS ON THE HEART AND THE CIRCULATION OF THE BLOOD.

Of what shape is your heart? — "My heart is shaped like a cone."

Where is it placed? — "In the chest, pointing toward my left side."

What bone is it near? — "It is near my breastbone."

Of what use is the heart? — "It contains the blood and sends it to the different parts of the body."

How much blood is sent from the heart at each beat? — "About two ounces."

What is the blood? — "A liquid made from food and drink."

Of what color is the blood? — "Bright red, when pure; dark red, when impure."

How does the heart send the blood through the body? — "Through pipes called arteries."

What do the arteries resemble in the way they are arranged? — "The branches of a tree."

What makes the blood impure? — "As the blood flows, it gives nourishment to every part of the body; this makes it poor. It also takes up the old worn-out particles; this makes it impure."

Where do the arteries carry the impure blood? — "To the veins."

Where do the veins carry the impure blood? — "To the heart."

Where does the heart carry the impure blood? — "To the lungs."

What happens to the impure blood in the lungs? — "It is made pure."

What makes it pure? — "Pure air."

Where do the lungs send the blood after it is made pure? — "Back to the heart."

Where does the heart send the pure blood? — "Throughout the body."

[48]

What is the journey of the blood to and from the heart to the different parts of the body called? — "The circulation of the blood."

What is the circulation of the blood? — "The circulation of the blood is its journey from the heart to the different parts of the body, and from the different parts of the body back to the heart."

How often does this circulation take place? — "Once in from three to eight minutes, according as the heart beats fast or slowly."

What kind of blood is necessary to health? — "Pure blood."

How should the blood circulate? — "Freely, all the time."

What do you mean by freely? — "Without anything to hinder."

What is necessary for the free circulation of the blood? — "I must wear clean clothing; I must exercise in work or play; I must keep my body warm."

How does tight clothing hinder the free circulation of the blood? — "By pressing upon the arteries and veins; and when about the waist, causing the ribs and other parts of the body to press upon the heart."

How does exercise help the free circulation of the blood? — "Exercise makes the heart beat faster, which causes the blood to more faster through the arteries and veins."

Why does keeping the body warm help the circulation of the blood? — "Because the blood moves faster when it is warmest; cold chills the blood, and makes it move slowly."

What harm do alcoholic liquors do to the heart? — "They make it tired, and sometimes wear it out."

In what way do they make it tired? — "They make it beat too fast."

Why does it beat too fast? — "Because it is hurrying to drive the alcohol out of the body."

In what other way do alcoholic liquors hurt the heart? — "They produce disease in it."

Tell one way by which the heart becomes diseased through [49] alcoholic liquors? — "Alcohol softens the fibres of the muscles of the heart, and fills them with fat."

What harm does this do to the heart? — "It makes it too weak to do its work, which is to pump the blood through the body."

What sometimes happens when the heart is thus weakened? — "It stops beating, which causes sudden death."

What harm does alcohol do to the blood? — "It uses up the water of the blood; it destroys the goodness of the red part; it makes the blood thin, impure, and unfit to do its work." See Appendices on Alcohol and Tobacco.

[50]

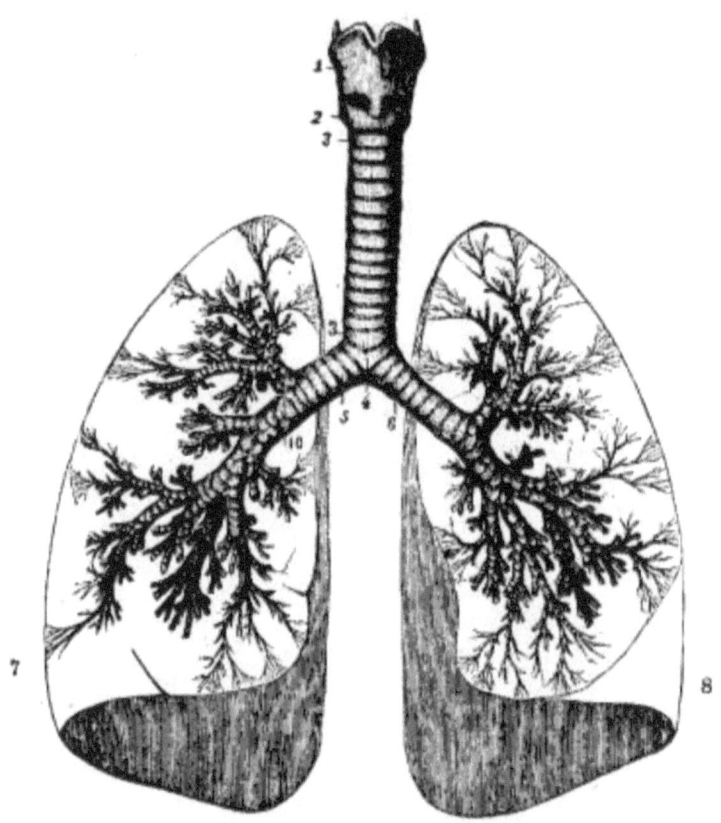

THE LUNGS.

    1, 2, the larynx, the upper part of the windpipe.

    3, the windpipe, or trachea.

    4, where the windpipe divides to right and left lungs.

    5, the right bronchial tube.

    6, the left bronchial tube.

    7, outline of the right lung.

    8, outline of the left lung.

    9, the left lung.

10, the right lung.

[51]

PART IX.

FORMULA FOR THE LESSON ON THE LUNGS AND RESPIRATION.

1. My lungs are the bellows or breathing machines of my body.

2. They are composed of a soft, fleshy substance, full of small air-cells and tubes. They are porous and spongy when healthy, but in some diseases become an almost solid mass, through which the air cannot pass.

3. I breathe by drawing the air through my windpipe, along the tubes into the cells of my lungs, swelling them out, and causing my chest to expand; then the chest contracts, and the impure vapor in my lungs is pressed out through the same tubes, windpipe, nose, and mouth, into the atmosphere.

4. I cannot live without breathing, because if the air does not go down into my lungs, the dark blood in them is not changed into pure red blood, and goes back through my body dark blood, which cannot keep me alive.

5. If I would have healthy lungs,

I must breathe pure air,

I must live in the sunlight,

I must keep my body clean,

I must wear loose clothing,

I must wear clean clothing,

I must sit and stand erect,

I must keep all parts of my body warm,

I must not change my winter clothing too early in the spring,

I must avoid draughts of cool air,

I must not rush into the cold when I am in a perspiration,

I must not poison my lungs with alcohol or tobacco.

QUESTIONS FOR THE FORMULA.

1. What are the lungs?
2. Describe the lungs.
3. How do you breathe?
4. Why can you not live without breathing?
5. What is necessary if you would have healthy lungs?

THE AIR AND THE LUNGS.

The air which enters through the nose and mouth passes into a tube of muscles and ring-like pieces of cartilage. The upper part of this tube is the voice-box or *larynx*, covered by a spoon-shaped lid which closes when we swallow; the lower part is the *trachea*, and the two parts are the windpipe. The trachea divides into two branches, *the bronchial tubes*, one for each lung. These tubes divide again and again like the branches of a tree, and end in exceedingly small sacs or bags. The air in these sacs, or air-cells, gives *oxygen* to the blood in the tiny blood-vessels of the lungs and takes from them the poison, *carbonic-acid gas*, water, and impurities, which it carries back through the windpipe into the outside air.

QUESTIONS ON THE LUNGS AND RESPIRATION.

Of what are the lungs composed? — "Of a soft, fleshy substance, full of small air-cells and tubes."

Of what use are the lungs? — "They are the breathing machines of the body."

How do the lungs appear when healthy? — "Porous and spongy."

How does the air get into the lungs? — "The air flows through the nose and mouth, into the windpipe and along the air-tubes, into the air-cells of the lungs."

[53]

What does the air do in the lungs? — "It swells the lungs and causes the chest to expand."

What do you mean by expand? — "To increase in size."

How is the air expelled from the lungs? — "The chest contracts and sends the impure air through the tubes and windpipe, the nose and mouth, into the atmosphere."

What do you mean by contracts? — "Becomes smaller."

What do you mean by atmosphere? — "The air."

Of what use is the air when it is in the lungs? — "It makes the blood pure."

Why can you not live without breathing? — "Because, if I do not breathe, pure air cannot get into the lungs to make the bad blood pure, and I cannot live if the dark, impure blood is sent back again through my body."

Why must you live in the sunlight? — "Because the sunlight helps to purify the blood and strengthen the body."

Why must you wear loose clothing? — "Because tight clothing stops the circulation of the blood."

Why must you avoid tight-lacing? — "Because tight-lacing crowds the ribs against the lungs, so that the lungs cannot move freely."

Why should you wear clean clothing? — "That nothing impure may pass into the body through the pores of the skin."

Why should you keep the body clean? — "That the pores of the skin may not be closed, but remain open to let the perspiration pass through."

What has the cleanliness of the body to do with the health of the lungs? — "If the body is not kept clean, the perspiratory pores become clogged."

What happens when the perspiratory pores are clogged? — "The impure particles which should pass through them stay in the body, and cause disease in the lungs or other parts."

Why should you sit and stand erect? — "Because, if I am in the habit of stooping, my lungs will be crowded, and will not have enough room to move freely."

Why should you keep all parts of the body warm? — "Because [54] chilling any part of the body causes the blood to chill in that part, and thus hinders its circulation."

Why should you not change your winter clothing too early in the spring of the year? — "I may take cold if not warmly clothed during the cool days of early spring."

Why should you avoid draughts of cool air? — "Because the cool air blows upon some parts of the body and closes the pores of the skin, checking the perspiration, and hindering the circulation of the blood."

Why should you not rush suddenly from a warm to a cool place? — "Because when warm the pores of the skin are open; if I rush suddenly into the cool air, these pores are closed too quickly."

Why does stopping the perspiration hurt the lungs more or less? — "The impurities it ought to carry away remain in the body, make the blood impure, and produce disease in some part; very often that part is the lungs."

What harm does alcohol do in the lungs? — "It fills the lungs with impure blood."

What harm does it do to the air-cells? — "It hardens the walls of the air-cells of the lungs."

What harm is done by the hardening of these air-cells? — "1. The lungs cannot take in enough of the gas called oxygen to purify the blood perfectly. 2. The gases or vapors in the lungs cannot pass freely through the hardened air-cells."

What happens from this? — "The lungs become diseased."

From what disease do some hard drinkers suffer? — "Alcoholic consumption, for which there is no cure." See Appendices on Alcohol and Tobacco.

[56]

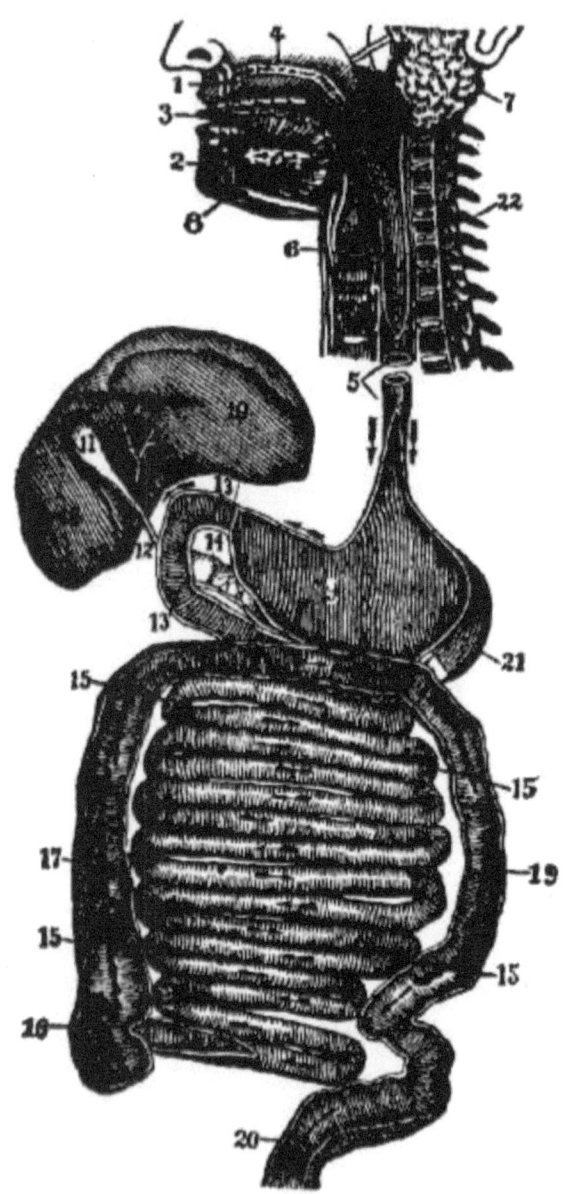

THE

DIGESTIVE ORGANS.

1. The upper jaw.
2. The lower jaw.
3. The tongue.
4. The roof of the mouth.
5. The food-pipe.
6. The windpipe.
7, 8. Where the saliva is made.
9. The stomach.
10. The liver.
11. Where the bile is made.
12. The duct through which the bile passes to the small intestine.
13. The upper part of the small intestine.
14. Where the pancreatic juice is made.
15. The small intestine.
16. The opening of the small into the large intestine.
17-20. The large intestine.
21. The spleen.
22. The spinal column.

[57]

PART X.

FORMULA FOR THE DIGESTIVE ORGANS AND DIGESTION.

1. When my food is chewed, it is rolled by my tongue into the oesophagus, or food-pipe, which is back of my windpipe, and leads from my mouth down along the side of my spine, to the left and upper end of my stomach.

2. My stomach is an oblong, soft, and fleshy bag, extending from my left to my right side, below my lungs and heart.

3. It is composed of three coats or membranes, and resembles tripe.

4. The *outer coat* is smooth, thick, and tough. It supports and strengthens the stomach.

5. The *middle coat* is fibrous. Its fibres have the power of contracting, sometimes pressing upon the food, and sometimes pushing it along toward the opening which leads out of the stomach.

6. The *inner coat* is soft, thick, spongy, and wrinkled. It prepares a slimy substance and a fluid. The slimy substance prevents the stomach from being irritated by the food. The fluid dissolves the food.

7. Food passes through several changes after it enters the mouth.

8. It is changed into pulp in the *mouth,* by the action of the teeth and the saliva. This is called *mastication*. It is changed in the *stomach*, by the action of the stomach and the gastric juice, into another kind of pulp called *chyme*. The chyme is changed by the bile and another kind of juice, called *pancreatic*[58]*juice*; these separate the nourishing from the waste substance. The nourishing, milk-like substance is called *chyle*. The waste substance passes from the body. The chyle is poured into a vein behind the collar bone, and passes through the heart to the lungs, where it is changed into blood.

9. If I would have a healthy stomach,

I must be careful what kind of food I eat,

I must be careful how much I eat,

I must be careful how I eat,

I must be careful when I eat.

10. I must eat wholesome food, good bread, ripe fruits, rather than rich pies or jellies.

11. I must eat enough food, but not too much.

12. I must eat slowly,

I must masticate my food thoroughly,

I must masticate and swallow ray food without drinking

13. I must take my food regularly but not too often,

I must rest before and after eating, if possible,

I must not eat just before bedtime.

14. I must breathe pure air,

I must sit, stand, and walk erect,

I must not drink alcoholic liquors,

I must not snuff, smoke, or chew tobacco.

QUESTIONS FOR THE FORMULA.

1. Describe the process of eating.[2] See page 21.
2. Where does the food go after it is chewed?
3. Describe the stomach.
4. Of what is the stomach composed?
5. Describe the outer coat of the stomach, and tell its use.
6. Describe the middle coat of the stomach, and tell its use.
7. Describe the inner coat of the stomach, and tell its use.
8. What happens to the food after it enters the mouth?
9. Tell about these changes.
10. What is necessary if you would have a healthy stomach?

11. What kind of food must you eat?

12. How much food must you eat?

13. How must you eat?

14. When must you eat?

15. What other rules must you obey?

[2] See Formula 7 on the Organs of Sense.

"EAT TO LIVE, NOT LIVE TO EAT."

There is pleasure in eating, because God has given us the sense of taste, that we may enjoy our food. But not everything which pleases this sense is good for the body, so we should learn what things are wholesome and choose them for our food and drink, refusing everything which is unwholesome. Those who obey these rules "*eat to live*" and never become drunkards or gluttons.

QUESTIONS ON THE DIGESTIVE ORGANS AND DIGESTION.

What happens to the food after it is chewed? — "It is rolled by my tongue into the oesophagus or food-pipe."

Where is the oesophagus or food-pipe? — "It passes from the mouth down the left side of the spine."

What is the stomach? — "A fleshy bag which receives and changes the food we eat."

Where is the stomach? — "In the front part of the chest, below the heart and lungs."

Of what is the stomach composed? — "Of three coats or membranes."

What do you mean by composed? — "Made of."

What do you mean by membrane? — "A thin skin."

What are the coats of the stomach called? — "The outer coat, the middle coat, the inner coat."

Describe the outer coat of the stomach. — "The outer coat is smooth, thick, and tough."

Of what use is the outer coat of the stomach? — "It strengthens and supports the stomach."

What do you mean by supports? — "Holds."

Describe the middle coat of the stomach. — "The middle coat is composed of fleshy fibres, which have the power of making themselves long or short."

What do you mean by fibrous? — "Composed of threads."

What do you mean by fibres? — "Threads."

Of what are the fibres of the stomach composed? — "Of flesh."

Of what use are the fibres of the stomach? — "They press upon the food, and push it toward the opening which leads out of the stomach."

Describe the inner coat of the stomach. — "The inner coat is soft, thick, spongy, and wrinkled."

Of what use is the inner coat of the stomach? — "It prepares a slimy substance and a fluid."

Of what use is the slimy substance? — "It prevents the stomach from being irritated by the food."

Of what use is the fluid? — "It dissolves the food."

What do you mean by slimy? — "Soft, moist, and sticky."

What do you mean by irritate? — "To produce unhealthy action."

What do you mean by dissolves? — "Melts."

Where is the food changed after it is taken into the mouth? — "First it is changed in the mouth; second, it is changed in the stomach; third, it is changed after leaving the stomach; fourth, it is changed in the lungs."

By what is it changed in the mouth? — "By the action of the teeth and the saliva."

By what is it changed in the stomach? — "By the action of the stomach and a kind of fluid called gastric juice."

By what is it changed after leaving the stomach? — "By the action of the bile and the pancreatic juice."

By what is it changed in the lungs? — "Nobody knows."

Into what is it changed in the mouth? — "Into pulp." [61]

Into what is it changed after leaving the stomach? — "Into chyle and waste substance."

Into what is it changed in the lungs? — "Into blood."

What is the change in the mouth called? — "Mastication, or chewing."

What is the change in the stomach called? — "Chymification, or chyme-making."

What is the change after leaving the stomach called? — "Chylification, or chyle-making."

What is necessary, if you would have a healthy stomach? — "I must be careful what kind of food I eat; how much I eat; and when I eat."

What kind of food must you eat? — "Wholesome food, etc." See Formula.

How much must you eat? — "Enough, but not too much."

How must you eat? — "Slowly."

How should your food be masticated? — "Thoroughly."

When must you eat? — "Regularly, but not too often."

When should you avoid eating? — "Just before bedtime."

What kind of air should you breathe? — "Pure air."

How should you sit, stand, and walk? — "Erect."

Why should you not eat too much food? — "Because, if I eat too much food, my stomach will have too much work to do in changing it into chyme."

Why should you eat slowly? — "That I may have time to masticate the food thoroughly."

Why should you masticate your food thoroughly? — "That it may be well prepared to enter the stomach."

Why should the food be well prepared to enter the stomach? — "Because, if it is not well prepared in the mouth, the stomach will have too much work to change it into chyme."

Why should you eat regularly, but not too often? — "Because the stomach needs rest, which it cannot have, if I eat too often."

Why should you avoid eating just before bedtime? — "Because, while I am asleep, the stomach cannot do the work of [62] changing the food as it ought to be changed; because the stomach should rest with the other parts of the body."

Why should you breathe pure air? — "Because pure air helps to make pure blood, which the stomach needs to make it strong and healthy."

Why should you sit, stand, and walk erect? — "That the stomach may not be crowded out of its place, or pressed upon by other parts of the body."

In what way does tobacco hurt the stomach? — "It poisons the saliva and prevents it from preparing the food to enter the stomach."

What harm does tobacco do inside the stomach? — "It weakens the stomach and makes it unfit to change the food into chyme."

How will wise children treat tobacco? — "Let it alone. They will not chew, snuff, or smoke the vile weed."

Is alcohol food or poison? — "It is poison."

How do we know it is not food? — "Because it cannot be changed into blood."

How has this been proved? — "Alcohol has been found in the brain, and other parts of drunkards, with the same smell and the same power to burn easily which it had when it was taken into the mouth."

How do you know it is a poison? — "Because it does harm to every part of the body, beginning in the stomach."

What harm does alcohol do in the stomach? — "It hinders the stomach from doing its work; it burns the coats of the stomach; it destroys the gastric juice; it hardens the food, so that it cannot be dissolved by the gastric juice."

What does the stomach do with alcohol? — "Drives it out as soon as possible."

Where does the stomach send it? — "Into the liver."

Where does the liver send it? — "To the heart; and the heart sends it to the lungs."

What do the lungs do with the alcohol? — "They drive it out as soon as they can."

[63]

Where do the lungs send some of it? — "Through the nose and mouth, into the air."

What harm does the alcohol do in the breath? — "It poisons the air; it tells that some kind of alcoholic liquor has been taken into the stomach."

From what you have learned about alcohol, what do you think is the only safe rule to obey concerning cider, beer, wine, and all alcoholic liquors? — "I must not drink them, if I wish to have a strong and healthy stomach."

[64]

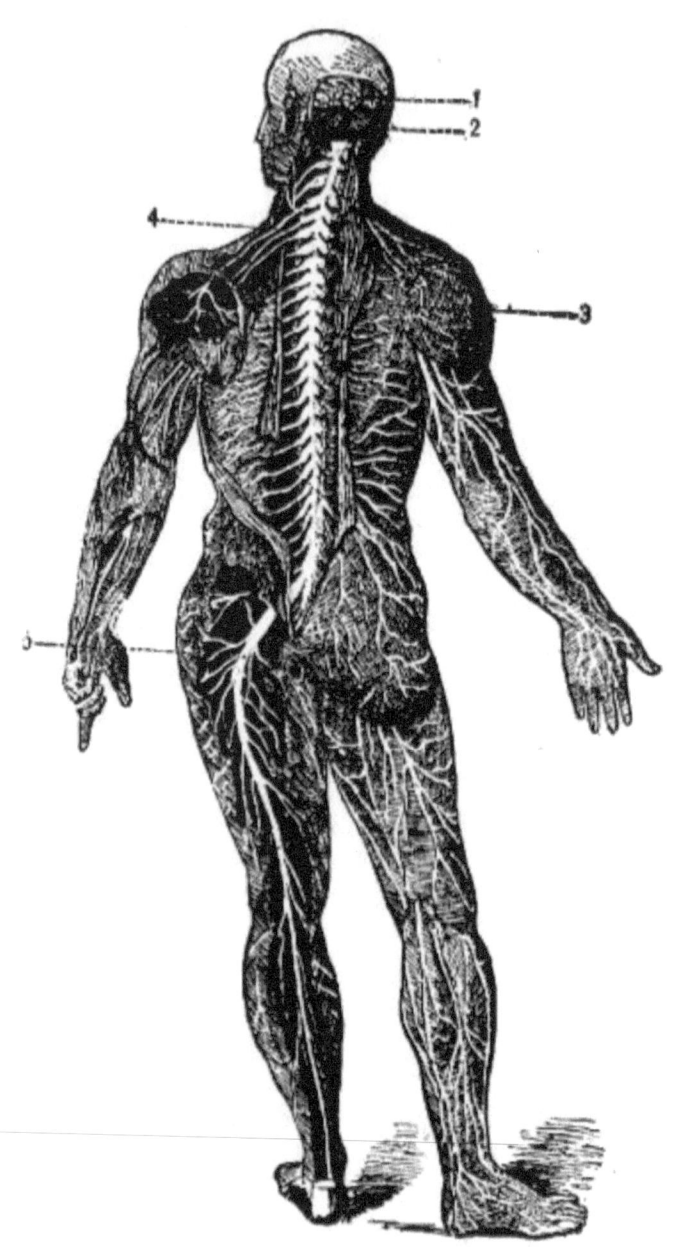

THE NERVOUS SYSTEM. — (From Walker's *Physiology*.)
1. The large brain. 2. The small brain. 3. The spinal cord. 4, 5. Nerves.

[65]

PART XI.

FORMULA FOR THE LESSON ON THE NERVOUS SYSTEM.

1. My brain is a soft gray-and-white mass resembling marrow.

2. It is placed in a bony box called the skull; it is covered and held together by three coats or membranes.

3. The outer membrane is thick and firm; it strengthens and supports the brain.

4. The middle membrane is thick, and somewhat like a spider's web in appearance.

5. The inner membrane is a network of blood-vessels.

6. From the brain, white or reddish gray pulpy cords, called nerves, pass to all parts of the body. These nerves are of two kinds: nerves of feeling, and nerves of motion.

7. If I prick my finger, a nerve of feeling carries the message to my brain; if I wish to move my finger, a nerve of motion causes my finger to obey my will.

8. Twelve pairs of nerves pass from the base of the brain: the first pair, called the nerves of smell, to my nose; the fourth pair, called the nerves of sight, to my eyes; the fifth pair, called the nerves of taste, to my mouth, tongue, and teeth. One pair pass to my face; another to my ears. The ninth, tenth, eleventh, and twelfth pairs to my tongue and parts of my throat and neck.[3]

9. The spinal cord is a bundle of nerves extending from the base of my brain, down through the whole length of my spine, or backbone. It is the largest nerve in my body.

10. From the spine, thirty-one pairs of nerves, called *spinal nerves*, pass to different parts of my body; some to the lungs, some to the heart, some to the stomach, some to the bones, and some to the muscles and skin.

11. If a nerve be destroyed it cannot carry messages to and from the brain. Before filling a tooth, the dentist sometimes destroys its nerve.

12. If a nerve be pressed upon too long it cannot perform its duty. If I press upon the nerve passing to my foot, I stop it from communicating with the brain; the foot loses its feeling, or, as I say, "is asleep."

13. If I drink alcoholic liquors, or snuff, smoke, or chew tobacco, my brain and nerves cannot do their work well; because alcohol and nicotine are very poisonous to the brain and nerves.

14. The brain must be supplied with good blood;

The brain must be used;

The brain must be rested;

I must drink wholesome drink, eat wholesome food, take enough exercise, and breathe pure air, that my brain may be supplied with pure blood;

I must study and think, that my brain may grow and be strong for work;

[67]

I must rest my brain when it is tired, either by changing my employment, or by going to sleep;

I must not poison my brain with alcohol or tobacco.

[3] NOTE.—*A fuller description of the Nerves of the Brain*: Twelve pairs of nerves pass from the base of the brain; the first pair, called the nerves of smell, to my nose; the second pair, called the nerves of sight, to my eyes; the third, fourth, and sixth pairs to the muscles of my eyes; the fifth pair to my forehead, eyes, nose, ears, tongue, teeth, and different parts of my face; the seventh pair to different parts of my face; the eighth pair, called the nerves of hearing, to the inner part of my ear; the ninth pair to my mouth, tongue, and throat; the twelfth pair to my tongue; the eleventh pair to my neck; the tenth pair to my neck, throat, lungs, stomach, and different parts of my body.

QUESTIONS ON THE FORMULA.

1. Describe the brain.
2. Where is the brain placed?
3. Describe the outer membrane of the brain.
4. Describe the middle membrane of the brain.
5. Describe the inner membrane of the brain.
6. Tell about the nerves.
7. Tell about the use of the two kinds of nerves.
8. Tell about the nerves which pass from the brain.
9. Tell about the spinal cord.
10. Tell about the nerves which pass from the spinal cord.
11. What happens if a nerve be destroyed?
12. What happens if a nerve be pressed upon too long?
13. What happens if you drink alcoholic liquors, or snuff, smoke, or chew tobacco?
14. What is necessary if you would have a healthy brain?

THE BRAIN AND ITS WORK.

The brain is egg-shaped, and of two parts, the large brain (*cerebrum*), and the little brain (*cerebellum*). These are composed of a white and gray substance, which in the large brain is so folded and wrinkled that it looks like the meat of an English walnut; in the little brain it is so arranged that it resembles a tree, and is called *arbor vitæ*, tree of life. The mind does its thinking through the large brain, and controls its muscles through the little brain.

A drunken man can not walk straight because alcohol has hurt the little brain; he can not think straight because it has poisoned the large brain.

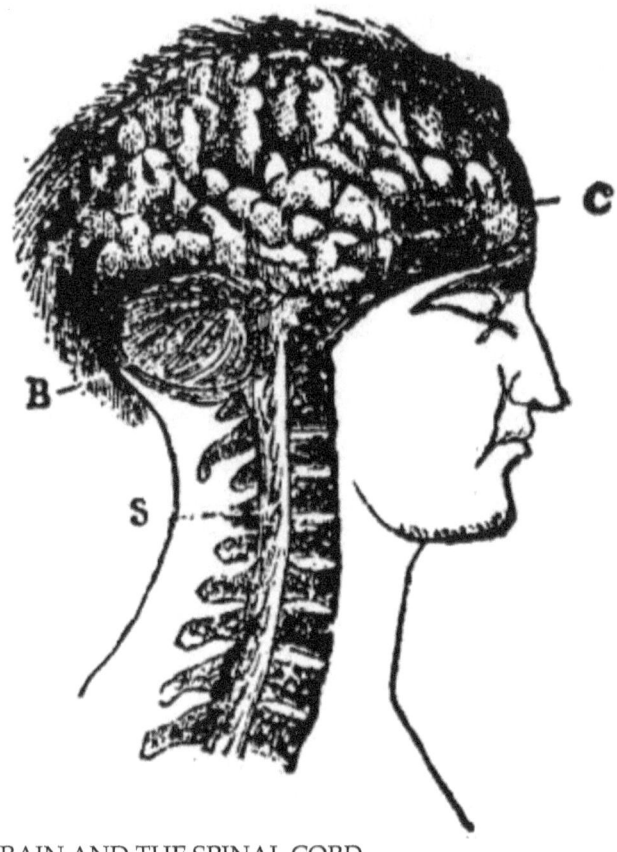

THE BRAIN AND THE SPINAL CORD.

C, the large brain (*cerebrum*). B, the small brain (*cerebellum*). S, a portion of the spinal cord.

QUESTIONS ON THE NERVOUS SYSTEM.

Where is your brain? — "In my skull."

What color is the brain? — "Gray and white."

What does the brain resemble? — "Marrow."

How is the brain protected? — "By three coats or membranes."

What may you name these membranes? — "The outer membrane, the middle membrane, and the inner membrane."

Describe the outer membrane. See Formula.

Describe the middle membrane. See Formula.

What are the nerves? — "White ashen-gray pulpy cords, which are found in the brain."

Where do they go from the brain? — "To every part of the body."

How many kinds of nerves have you? — "Two."

What names are given to the two kinds of nerves? — "Nerves of motion and nerves of feeling."

Which is the largest nerve in the body? — "The spinal cord."

[69]

Where is the spinal cord? — "It extends from the brain throughout the whole length of the backbone."

How may you describe the spinal cord? — "It is a bundle of nerves, etc." See Formula.

Where are the spinal nerves? — "They pass from the spinal cord to different parts of the trunk and limbs."

How many pairs of nerves pass from the base of the brain? — "Twelve."

Where do the first pair go? — "To the nose."

What are they called? — "The nerves of smell."

Where do the second pair go? — "To the eyes."

What are the second pair called? — "The nerves of sight."

Which move the muscles of the eyes? — "The third, fourth, and sixth pairs."

Where do the fifth pair go? — "To the forehead, eyes, nose, ears, tongue, teeth, and different parts of the face."

The seventh pair? — "To the different parts of the face."

The eighth pair? — "To the inner ear."

What are the eighth pair called? — "The nerves of hearing."

Where do the ninth pair go? — "To the mouth, tongue, and throat."

Where do the twelfth pair go? — "To the tongue."

Where do the eleventh pair go? — "To the neck."

Where do the tenth pair go? — "To the neck, throat, lungs, stomach, and different parts of the body."

What happens if a nerve be destroyed? — "It cannot carry messages to the brain."

What happens if a nerve be pressed upon too long? — "It cannot carry messages to the brain."

What is necessary if you would have a strong, healthy brain? — "My brain must be used; my brain must be rested; my brain must be supplied with pure blood."

How must you use your brain? — "In thinking and studying."

How may the brain be rested? — "By sleep."

In what other way may the brain be rested? — "By thinking of something different from that which made it tired."

[70]

What two brain-poisons have you learned about? — "Alcohol and tobacco."[4]

With what may you show the harm done by alcohol to the gray part of the brain? — "With alcohol and the white of an egg."

How could you show it with these? — "I would pour the alcohol upon the white of the egg."

What would then happen? — "The white of the egg would harden as if it had been boiled."

What is in the white of an egg? — "Water and albumen."

Where else may we find albumen? — "In some seeds, and in the gray part of the brain and the nerves."

What harm does alcohol do to the nerves? — "It takes away their moisture and hardens them."

What harm does this do to them? — "It paralyzes them, or makes them lose their power."

What happens when nerves are paralyzed? — "They lose their power over the muscles; they are unfit to carry messages to and from the brain."

What harm does alcohol do to the gray part of the brain? — "It hardens it, as it hardens the white of an egg."

What harm does this do to the brain? — "It paralyzes it, or makes it lose its power."

What then happens? — "It cannot properly do its work of thinking, and cannot control the nerves."

What disease is sometimes caused by this hardening of the brain by alcohol? — "Paralysis, which often ends in death."

What harm does alcohol do to the blood-vessels of the brain? — "It fills them with impure blood."

What disease is caused by the blood-vessels of the brain being filled with impure blood? — "Congestion of the brain, or apoplexy, which ends in death."

What else frequently happens to those who drink alcoholic liquors? — "They become crazy, or insane."

[71]

If you wish to have a strong, healthy brain, what should you do about these liquors? —

"Never put them into my mouth,

To steal away my brains."

Tell of what dreadful disease people die who are bitten by a mad dog. — "Of hydrophobia."

Of what dreadful disease do people sometimes die who are bitten by the serpent in alcoholic liquors? — "Of delirium tremens."

Which is the more dreadful, hydrophobia or delirium tremens? — "One is as dreadful as the other."

How can you be sure never to have delirium tremens? — "By drinking nothing which has alcohol in it."

Will a little beer or wine hurt you? — "Yes, it may make me love the taste of alcohol."

What harm is there in loving the taste of alcohol? — "I may love it so much as to become a drunkard."

Tell once more how you should treat alcoholic liquors. — "I should never drink a drop of them."

[4]See Appendices.

[73]

ALCOHOL.

THE STORY ABOUT ALCOHOL.

Several hundred years ago many people were trying to discover something that would keep them young and strong, and prevent them from dying. It is said by some that a man named Paracelsus, in making experiments, discovered *alcohol*. He called it "the water of life," and boasted that he would never be weak and never die; so he went on drinking alcoholic liquors until at last he died in a drunken fit.

What is this alcohol which has done and is doing so much mischief in the world? I will show you some. What does it look like? — "Water." Yes; and if you were to smell it you would say it has a somewhat pleasant odor; if you were to taste it, that it has a hot, biting taste, *i.e.*, is pungent. If you put a lighted match to it you would notice that it burns easily, and with a flame, and may therefore be said to be combustible and inflammable.

What does it come from? Is it one of the drinks God has given us? Some of the class think it is; we will try to learn whether this answer is correct or not. If we study about it very carefully we shall discover that it is not a natural drink, that it is not found except where it has been made from decayed or rotten fruits, grains, or vegetables.

If you take some apples, and squeeze the juice out of them, you will find it sweet and pleasant; let that juice stand for several days and what will happen to it? — "It will get bad." Yes; or, as grown people say, it will *work* or *ferment*; that is, the sugary part of the juice will be separated into a kind of gas and a liquid. The gas is called *carbonic acid gas*; the liquid is *alcohol*. Both the gas and the liquid are poisonous.

Alcohol may also be obtained from other fruits, as grapes, and from some grains and vegetables. But all these must first become rotten before alcohol will come out of them. This is [74] one reason why we think that God, who gives us good, wholesome food, did not intend alcohol to be a drink for man, else He would have put it into the delicious ripe fruit, and not made it impossible to get until they decay.

Now let us put upon the blackboard something which will help us remember what we have learned about

ALCOHOL.

DISCOVERED BY	DESCRIPTION.	MADE FROM
Paracelsus.	Water-like; with a pleasant odor; a hot, biting taste; and will burn with a flame.	Fruits, Grains, or Vegetables.
CALLED "The water of life."		

USES OF ALCOHOL.

We put some sugar into water; the children see that it melts; then some glue or shellac is placed in the same liquid; they see that this is not melted, but that, when alcohol is used instead of water, the glue or shellac is dissolved. From this experiment they learn that alcohol is used in making varnishes.

Some water is poured into one saucer, and alcohol into another; a lighted match is applied to each; the class notices that the alcohol takes fire and burns, while the water does not.

Next, we fill a lamp with alcohol, and put a wick into it; when the wick becomes wet with the fluid it burns steadily and without smoke, as may be seen by holding a clean white saucer over the flame. This shows why jewellers and others, who wish to use a lamp to make things very hot, prefer alcohol to kerosene, which, as the children know, smokes lamp-chimneys, or anything else, so easily.

We show a thermometer; the children are told its use if they are not already familiar with the instrument; we talk about the quicksilver in the tube, about its rising or falling according to the degree of heat or cold; then we inform the class that in [75] some countries where it is very cold quicksilver freezes; for this reason alcohol, which does not freeze, is colored red and put into the thermometer tube to be used in these Arctic regions.

Another use for alcohol is to keep or preserve substances. This we illustrate by placing a piece of meat into some alcohol. We explain that the water in the meat is that which causes it to decay. Alcohol has the power to take up or *absorb* water; so when meat is put into this liquid the water from the meat is absorbed by it, and the meat does not become bad. Those who wish to preserve insects a long time, and doctors who desire to keep any portion of a human body after death, put these into alcohol, in which they may be kept for a long time.

Lastly, we let the children smell cologne or other perfumery, and tell them this is made from different oils mixed with alcohol.

At the close of this lesson the class is ready to help us make the following BLACKBOARD OUTLINE.

FACTS ABOUT ALCOHOL.	GOOD USES OF ALCOHOL.
It melts gums.	To melt gums.
Burns with a flame.	To make varnishes.
Burns without smoke.	To burn in lamps.
Will not freeze.	To make camphene, etc.
Likes water.	To put into thermometer tubes.
Mixes with oils.	To preserve meats, etc.
	To make perfumery.
	In making jewelry.

USES OF ALCOHOL—*concluded*.

You see alcohol is very useful for some purposes; but do people ever drink it? Some of the children think not, and we grant that no one is foolish enough to drink *raw* alcohol, because it is too strong. It would take only a little to make them drunk, and only a few ounces to kill them instantly.

We ask the pupils if they have ever seen a drunken person, and what made that person drunk? We soon obtain an answer, [76] and place upon the board "Rum, gin, whiskey, brandy," as the names of drinks which will take away the good sense of those who drink them. To these are added "Wine, beer, ale, lager, and cider."

We explain that all these have alcohol in them, as may be known by smelling them, or by smelling the breath of those who have drunk even a little of them; and that because they contain alcohol they are called *alcoholic liquors*. If a person drinks any one of them he will be poisoned, more or less, according to how much he takes. The children are astonished at the word *poisoned*, but we explain that the very word, *intoxicated*, means poisoned. So a drunken man is a poisoned man. If enough alcohol, or alcoholic liquor, is drunk by anyone, he will drop down dead as quickly as if he were shot by a cannon ball.

When told that alcohol is not a food, but a poison, the class readily understands what we mean, and we have no difficulty in having the following statements prepared and memorized:

FOOD.

That which makes the body grow, and helps to keep it alive.

POISON.

That which hurts the body, and makes it die.

ALCOHOL.

QUALITIES.	GOOD USES.
Water-like, *looks like water*.	To melt gums.
Transparent, *may be seen through clearly*.	To make varnishes.
	To burn in lamps.
Odorous, *has a smell*.	To make camphene, etc.
Pungent, *has a hot, biting taste*.	To put in thermometer tubes.
Liquid, *will flow in drops*.	To preserve meats, insects, etc.
Poisonous, *hurts the body*.	To make perfumery.
Intoxicating, *takes away the senses; makes drunk*.	In making jewelry.
Absorbent, *takes up or absorbs water*.	BAD USE.
	To drink.
Inflammable, *burns with a flame*.	
Uncongealable, *will not freeze*.	
Innutritious, *not good for food*.	

ABOUT FERMENTATION AND FERMENTED LIQUOR.

Alcohol.—Alcohol may be obtained from any substance which contains sugar or starch, or both sugar and starch, as apples, pears, grapes, potatoes, beets, rice, barley, maple, honey, etc.

Alcohol can be obtained only by *fermentation*. By fermentation we mean the change which takes place when a juice containing sugar decays, or goes to pieces. You know decay always makes things fall to pieces.

You ask, what pieces is sugar made of? Very, very little pieces, called *atoms*. There are different kinds of sugar. In that made from grapes, called *grape sugar*, there are six atoms of carbon, twelve of hydrogen, and six of oxygen. What are carbon, hydrogen, and oxygen? Oxygen is the kind of gas which keeps animals alive, and makes things burn. Hydrogen is another kind, which you have smelled perhaps when water has been spilled on a hot stove; the gas burned in street-lamps is hydrogen that has been driven out of coal. Carbon you see in charcoal and soot; the black lead of your lead-pencils is mostly composed of carbon and iron; lamp-black is pure carbon, without form or shape.

We will let these circles of colored paper stand for the atoms of carbon, hydrogen, and oxygen in grape sugar,—the largest, which are red, for the oxygen; the second size, which you notice are black, will represent atoms of carbon; while the little blue ones will make you think of hydrogen.

If you remember that it takes one atom of carbon and two of oxygen to make carbonic acid gas; also, that two atoms of carbon, one of oxygen, and six of hydrogen to form alcohol, you can easily find that two atoms of carbonic acid gas and two atoms of alcohol may be formed from an atom of sugar. So the more sugar a juice contains the more alcohol may be formed from it.

Cider.—Cider is made by pressing the juice out of apples. This sweet cider ferments, and the sugar part of it changes into carbonic acid gas and alcohol. People who do not understand this go on drinking cider, not knowing that it makes drunkards of those who drink much of a beverage which seems so pleasant and harmless.

Wines. — Wines are made from the juices of fruits which have sugar in them, especially grapes. Sometimes people have what they call *home-made wines*, which they make from blackberries, currants, elderberries, gooseberries, cherries, or other fruits. They may ask you to take some, saying, "This will do you no harm; we did not put any alcohol into it." They do not know what you have learned, that alcohol is always formed in fermented juices which contain sugar. It does not wait to be put into the home-made wines; it quietly comes in as they are getting made, at home or any other place, and will make people drunk as surely as when it is found in brandy or any other liquor.

Some of the wines in the stores are made from grape juice, but many more are made by mixing hurtful and poisonous things together to make the liquor strong, and give it what is called a fine color and good taste.

Beer and Ales. — These are made from grains and hops, which contain no sugar, it is true, but are composed of starch, which may be changed into sugar. When a seed of grain is put into the ground and begins to grow, the starch in it becomes sugar, which feeds the young plant. When a brewer wishes to make beer, he takes some grain, puts it in a dark place, wets it, and leaves it to sprout, or begin to grow. Then he puts it into an oven to dry it, and make it stop growing. This makes what is called *malt*. The malt is mashed and soaked in warm water to get the sugar out of it; this forms a liquid called *sweet wort*. The wort is separated from the mashed grain and boiled; yeast is mixed with it to help it to ferment more quickly; it soon becomes changed; a dirty yellow scum filled with bubbles comes to the top, which we know is the poisonous carbonic acid gas; [79] the other poison, alcohol, stays in the liquid and makes the beer taste good to those who like it.

Liquors made from grain are called *malt liquors*. Lager beer, and all kinds of ales and porters, are malt liquors. They make people dull, sluggish, and stupid who drink much of them. They do much mischief in the body, though it takes a larger quantity of any one of them to make a person drunk than it does of whiskey or brandy.

AN ATOM OF

GRAPE SUGAR.	CARBONIC ACID GAS.	ALCOHOL.
Carbon, 6 atoms.	Carbon, 1 atom.	Carbon, 2 atoms.
Oxygen, 6 atoms.	Oxygen, 2 atoms.	Oxygen, 1 atom.
Hydrogen, 12 atoms.		Hydrogen, 6 atoms.

SUB-FERMENTED GRAPE SUGAR MAKES 2 atoms of carbonic acid gas and 2 atoms of alcohol.

ALCOHOLIC LIQUORS MADE FROM

FRUITS.		GRAINS.
Cider.	*Wines.*	*Beer, Ales, etc.*
Apples.	Grapes, Gooseberries, Barley,	Oats,
Perry.	Currants, Elderberries, Wheat,	Peas, etc.
Pears.	Blackberries, Cherries, etc. Corn,	(with hops).

DISTILLATION.

How does the sugar in grapes and other fruits become alcohol? — "By fermenting." Yes, and liquors made by fermenting are called *fermented liquors*. What other alcoholic drinks have you heard about beside cider, wines, beer, and ales? — "Gin, whiskey, brandy, rum." These are stronger than the fermented liquors, that is, they contain more alcohol; they are made by what is called *distillation*.

If you boil water, and let the steam from it fall upon a cold plate, the steam will change back into liquid and become [80]*distilled* water. Making a liquid boil, catching the vapor or steam and cooling it, is what we mean by distillation.

If two or more liquids are mixed together, the one that boils with the least heat will be drawn off first. The alcohol of beer, cider, and wines is mixed with water; it boils at a lower heat than water, so can be drawn off from it very easily. This does not make more alcohol, it only makes the alcohol stronger by separating it from the water.

When beer or any other alcoholic liquor is to be distilled, it is poured into a large copper boiler, called a *still*, and boiled. A tube carries the vapor from the boiler into a cask filled with cold water. This tube is coiled like a spiral line or worm through the cask; it is called *the worm of the still*, and the cask is *the worm-tub*. As the vapor passes through the tube, it cools and drops out at the end into the worm-tub, changed into a liquid stronger in alcohol than that from which it was drawn or distilled.

In this way gin is made from beer, brandy from wine, and rum from fermented molasses. These are very strong drinks, and only hard drinkers like them. But very few people begin by taking these; they first learn to like alcohol by drinking cider, beer, or wine, and end with gin, whiskey, or rum when they have become drunkards.

DEFINITIONS.

Distillation. Drawing the vapor from a boiling liquid and cooling it.

Still. Machinery for distilling; the boiler which holds the liquid.

The Worm of the Still. The tube which passes from the still to a cask, in which it coils like a worm.

Worm-tub. The cask which holds the tube or worm, and receives the distilled liquid.

Distilled Liquid. A liquid formed by cooled steam.

Distilled Liquors. Liquors made by distilling alcoholic liquors.

Fermented. Changed by decay.

Fermented Liquors. Liquors which have been fermented or changed by decay, and contain alcohol.

Unfermented. Not decayed.

Unfermented Liquors. Liquors which contain no alcohol.

[81]

KINDS OF LIQUORS

[5]UNFERMENTED.	FERMENTED.	DISTILLED.
Grape juice,	Hard cider,	Gin,
Sweet cider,	(Malt liquors)	Brandy,
Root beer,	Beer,	Whiskey,

Ginger beer.   Lager beer,   Rum.
Perry.   Ale,
                Porter,

                Wine.

[5] These soon become fermented; they then contain alcohol.

HARM DONE BY ALCOHOL IN VARIOUS PARTS OF THE BODY.

Raw alcohol does not do much harm to people because it is too strong for them to drink much of it; but the alcohol hidden in cider, ale, wine, whiskey, and other alcoholic drinks kills not less than *sixty thousand* persons in this country every year, besides those who die from its use in other parts of the world.

There is great excitement when there is a mad dog around; and, if any one is bitten and dies from the dreadful hydrophobia, people are ready to destroy all the dogs of the neighborhood; but when a drunkard dies from delirium tremens or alcohol craziness, how few take any notice of the cause of his death, or do all they can to wage war against the use of alcoholic liquors.

But why do we say such hard things against these liquors which some people love so well and think so harmless? In what way do they hurt and kill people? Let us see. Where does what we drink go after it has been put into the mouth?—"Into the stomach." If it were the right thing to go into the stomach, into what would it be changed?—"Into something which helps to make good blood."

Learned men, who have examined and carefully studied about these things, tell us that *the stomach is hurt* by alcohol, [82] because the fiery fluid is not food, but poison which makes the stomach very sore, and gives it hard work to do. The veins of the stomach take it up and send it into the liver. The liver, which is a large organ weighing about four pounds, lies on the right side below the lungs; its work is, to help make the blood pure. It can do nothing with alcohol, so it drives it along to the heart; the heart sends it to the lungs; the lungs throw some of it out through the breath, which smells of the vile stuff that has been poisoning every part it has passed through since it entered the mouth.

Some of the alcohol does not get out of the lungs through the breath, but goes with the blood back to the heart, and from the heart is sent through the arteries to every part of the body. No part of the body wants it.

The Skin drives some of it out, through its little pores, with the perspiration.

The Kidneys, which lie in the back below the waist, on each side of the spine, send off some of the poison.

Yet some of it gets into *the brain*, and there does very much mischief, of which you will learn more by and by. You know, if the brain is hurt, the mind cannot do its work of thinking properly; thus, alcohol does great *harm to the mind* through the brain.

The muscles and *the bones* are hurt by not being supplied with pure blood; *the heart* gets tired out with overwork, and *the lungs* become diseased through this same terrible alcohol.

Therefore, if you would be strong and healthy, have nothing to do with alcoholic liquors; for

ALCOHOL POISONS

The stomach,	The liver,	The blood,
The heart,	The lungs,	The brain,
The bones,	The muscles,	The skin,

And every part of the body.

[83]

IN THE STOMACH.

Children who have learned the Lesson on Digestion, and know about the coats of the stomach, about mastication and chyme-making, are easily made to understand why anything which has alcohol in it is unfit to go into the stomach.

If we touch a drop of alcohol to the eye, it will make it sore; so alcohol in the stomach irritates its coats and makes them sore.

Alcohol poisons the gastric juice. If we get some of this juice from the stomach of a calf which has just been killed, and mix alcohol with it, the alcohol will separate the watery part from the *pepsin* or white part. This is what alcohol does in the stomach. It takes up water from the gastric juice, which prevents the pepsin from mixing well with the food, and hinders the change of the food into chyme, which cannot take place without pepsin.

The children have already learned that alcohol keeps meat from decaying, or going to pieces. We explain that food in the stomach must go to pieces to prepare it to make blood; when mixed with alcohol, it is preserved, and the gastric juice cannot melt or dissolve it. Thus the stomach is hindered from doing its work until it gets rid of the alcohol.

A true story we have read will help you to remember how troublesome alcohol is to the stomach. Some men in Edinburgh were paid their wages, one Saturday, soon after they had eaten their dinner. They got drunk and remained so till the next day at noon. When they became sober they had a headache and were so ill that they sent for a doctor; he gave them some medicine which brought up their Saturday's dinner just as it had gone down into the stomach. The poor stomach could do nothing with dinner mixed with whiskey or rum, because these liquors are half alcohol.

You have already learned that the stomach hurries to drive out the alcohol into the liver; the liver sends it with the blood into the heart; the heart pours it into the lungs; the lungs breathe it out through the nose and mouth, and tell that some kind of alcoholic liquor has been taken into the stomach.

[84]

Remember, that the alcohol which comes out in the breath is a part of that which *went into the mouth*. It could not be changed. It did nothing but mischief in its journey, which shows that it is not food, but poison. God, who created the body, has not given any part of it power to change alcohol into blood.

People sometimes take ale or wine because they think it gives them an appetite. This is a great mistake. When any alcoholic liquor goes into the stomach, there is such hard work to get it out that the pain of hunger is not felt; when it is out, the stomach is tired and does not tell the brain that it is hungry. When alcohol is poured into it, day after day, it loses its desire for good, wholesome food, *and wants more and more alcoholic liquor*. It has an appetite for alcohol.

Alcohol makes the stomach sore and full of disease; people who take much of it in liquors always suffer much from dyspepsia.

So, if the stomach could speak, it would say: "Don't pour any alcohol into me, though you mix it and call it ale, cider, wine, or any other name that makes folks think it will do me no harm. You cannot deceive me. I know alcohol as soon as it comes down, and it always makes me suffer."

BLACKBOARD OUTLINE.

ALCOHOL—
    Burns or inflames the coats of the stomach.
    Spoils the gastric juice.
    Makes the food hard to be dissolved.
    Makes the stomach tired and weak.
    Takes away the appetite for wholesome food.
    Makes an appetite for alcoholic liquors.
    Causes disease in the stomach and other digestive organs.

QUESTION ON BLACKBOARD OUTLINE.

What harm does alcohol do in the stomach?

[85]

TO THE BONES, MUSCLES, AND SKIN.

To the Bones. — You have already learned that the bones require to be supplied with good blood to make them strong and healthy, and that alcohol does not make good blood, so we need spend no time in deciding that alcoholic liquors do injury to the bones, and that the bones of those who drink these liquors are less likely to heal, when broken, than those of persons whose blood has not been poisoned by alcohol.

To the Muscles. — The muscles, as you know, cover and move the bones; good blood makes them grow, and keeps them healthy and strong. People like to have plenty of good muscle, for this not only gives them strength, but makes them look plump and well.

Alcohol poisons the blood by killing many of the very little, round, red parts in it, called by a long name, which you can learn if you try. This hard name is *corpuscles* [kor'pussls]; *corpuscle* means *a little body*.

These little bodies float in the fluid portion of the blood, and go to every part of the body to help keep it alive and healthy. When alcohol hurts them, they turn into a poor kind of fat, like suet, and cannot do any good. They stay in different parts and do much harm. Sometimes they lodge between the muscles, and make a person look strong because plump; but he is not strong, for his muscles are filled with fat.

Sometimes the liver or the heart, which are only large muscles, become so heavy and soft with fat that they cannot do their work properly; they become weak and diseased, wear out, and cause the death of their owner, who has poisoned them with ale, wine, or other alcoholic drink.

To the Skin. — Alcohol hurts the skin also, by feeding it with poisoned blood, by giving the pores extra work in carrying off some of the alcohol in the perspiration, and by making the little blood-vessels larger than they should be in a way you will learn more about by and by. These little blood-vessels become very full of blood, and cause the red face and blue nose which [86] mark the drinker of alcoholic liquors. This redness of the skin tells of the mischief which alcohol is doing inside of the body. It is the danger-signal which warns against the use of the fiery poison.

ALCOHOL HURTS

THE BONES,	THE MUSCLES,	THE SKIN,
By supplying them with bad blood.	By supplying them with bad blood; By loading them with fat which makes them weak.	By supplying it with bad blood; By over-working the perspiratory pores.

TO THE BLOOD, THE LUNGS, AND THE HEART.

To the Blood. — The wonderful fluid which is the life of the body consists of a water-like liquid in which floats millions of the very little, circle-shaped, red particles which you have been taught to call *corpuscles*. You have also been told that alcohol kills these little bodies, and thus takes some of the life out of the blood, and fills it with useless, suet-like fat.

The blood, you know, flows everywhere through the body, giving its goodness to make every part grow and live, and carrying away the worn-out particles it meets. Blood, when poisoned with alcohol, goes through the body, giving disease and death instead of health and life. So, if you want good, red blood, do not let alcohol get into it.

To the Heart. — When alcohol comes with the blood from the liver, the heart begins to beat fast to get rid of the firewater; this makes it very tired, for it always has enough to do in carrying bad blood to the lungs, and pumping good blood into the arteries, without having the extra trouble of driving out alcohol. Wise people will not give it this extra work to do.

Besides, we told you, in the talk about the harm done by alcohol to the muscles, that the heart, — which is only a large [87] muscle, or rather many muscles fastened together so as to make a pear-shaped organ about the size of your fist, — is hurt in another way by alcohol. It gets too much of the poor kind of fat from the blood, which fills between the muscles, and after awhile makes the walls of the heart so soft and weak, that we could almost push through them with a finger, if we could get at them.

Very often the tired, overworked, weakened heart suddenly stops beating, and the person who would keep on drinking beer, wine, brandy, or rum falls down dead. "Died from heart disease," people say, when the truth is, *died from drinking alcoholic liquors*.

To the Lungs. — What are the lungs? — "The breathing-machines of the body." What do they throw out? — "Bad air." What do they take in? — "Fresh air." In pure air there is a good kind of gas which is necessary to keep us alive; this gas is called *oxygen*.

When air is taken into the lungs, the oxygen mixes with the blood in them and makes it pure. If alcohol is in the lungs, it hardens the walls of their air-cells, and keeps out the oxygen or good gas; at the same time it keeps in the impure gas, called *nitrogen*, which ought to come out through the nose and mouth into the air. Thus the blood in the lungs cannot be properly purified, and goes back to the heart impure blood which is unfit to be used.

The lungs are also obliged to work faster when alcohol is in them, because with the heart they are striving to drive out the enemy. This makes the lungs tired, sore, and inflamed. They are not as strong to do their work, and are more likely to breathe in any contagious disease than are the lungs of people who do not drink alcoholic liquors.

Some people go on drinking these poisons for many years, and seem not to be hurt by them; but at last they suffer from what is called Alcoholic Phthisis, a kind of consumption which doctors cannot cure.

[88]

HARM DONE BY ALCOHOL TO THE

HEART.	BLOOD-VESSELS.	LUNGS.
Overworks it.	Hurries the blood through them.	Makes them work too fast.
Makes it tired.		
Loads it with fat.	Stretches the small arteries and makes them unfit to work.	Heats and inflames them.
Softens and destroys it.		Hardens the walls

Poisons the blood in the hair-like blood-vessels (capillaries).	of their air-cells. Keeps in the poisonous gas. Keeps out the good gas (oxygen). Weakens them and makes them diseased.

THE BLOOD ("The life ... is in the blood")

Consists of

    A colorless liquid (plasma), and
    Little, red, circle-shaped bodies (corpuscles).

ALCOHOL (a blood-poison)

Mixes with the colorless liquid, and takes away some of its goodness.

Makes some of the corpuscles

    Smaller.
    Change shape.
    Lose color.
    Lose oxygen.
    Die, and change into useless fat

TO THE BRAIN AND NERVES.

Where is your brain? — "In my skull." What color is it? — "Gray and white." What does it resemble? — "Marrow." What work is done in the brain? — "The work of thinking." You may repeat what you have learned about the membranes of the brain. (See Formula for the Lesson on the Nervous System.)

[89]

You say "the inner membrane is a net-work of blood-vessels." If these are blood-vessels in the membranes, what fills them?—"Blood." Do you think alcohol can get into the brain?—"Yes." How can it get there?—"It goes there with the blood." How can we know that alcohol does mischief in the brain? You cannot answer? Did you never see a drunken man? Now tell me how you might know his brain has been hurt by alcohol.—"He talks funny; he acts strangely; he is very cross; he does not know what he is doing; he walks crookedly; he falls down; sometimes he falls asleep, and is almost like a dead man; he is dead drunk."

Let us study to learn why the drunken man does such strange things. The alcohol in this bottle, and this egg which you see, will help us find the cause of the mischief. You may tell what is in the egg.—"A white liquid and a yellow liquid." How could they be made hard?—"By making the egg hot; by boiling." We will try what alcohol will do to the white part. You see when it is poured upon the white of the egg it hardens this part as boiling would harden it. This white portion is composed of water and something called *albumen*. The alcohol dries up the water and thickens the albumen.

Albumen is found not only in eggs but in some seeds, as beans, peas, corn, etc., also in the gray part of the brain and in the nerves.

We will talk first of the harm alcohol does to the nerves. You know they are the grayish-white cords which pass from the brain and the spine to every part of the body. What do they act like in the kind of work they do?—"Like telegraph wires." What is their work?—"To carry messages to and from the brain." What kinds of nerves have you learned about?—"Nerves of feeling and nerves of motion."

When alcohol touches a nerve, it draws away the moisture or water from it, and hardens the white part or albumen; this makes the nerve shrivel as if it had been burned; it loses its power to feel and move, or, to use a long word, is *paralyzed*.

Alcohol paralyzes all the nerves it touches. It makes them [90] so stupid that they cannot understand what the brain says to them, and they do not carry the right messages back to it. For instance: when the nerves of the stomach are poisoned by the alcohol in beer, wine, etc., they do not feel the pain of hunger as much as they otherwise would, and they let the brain think the stomach is satisfied and does not need any more food, when it is only stupefied by these liquors.

Again, it is the work of some nerves to tell the muscles of the small arteries to tighten, or contract, when too much blood is coming into them. Alcohol so paralyzes these nerves that they do not carry their message; the arteries let in the blood, and become swollen and enlarged. They tell the mischief done to them, by causing the skin to be red or flushed. If people drink much of any intoxicating liquor, and often, their skin is always a bad color, or, as grown folks say, becomes permanently discolored. All this because the nerves have been made unfit to do their duty by alcohol poison.

The nerves also lose power over the muscles of the limbs. This is plainly seen in the trembling of the hands and the unsteady walking of the drunkard; but is equally true of those who drink only a little now and then. Their nerves are not as strong and wide-awake to control the machinery of the body as they would be if no alcohol were troubling them.

Sometimes the nerves of hearing and sight tell the brain queer stories, and the poor brain believes them all, for it, too, is stupefied by the same fire-water which has hurt the nerves. Indeed, the harm done by alcohol to the brain is greater than that done to any other part of the body. It takes the water from the albumen, and makes the white part of the brain hard, as if it had been cooked. It kills the little, circle-shaped, red parts of the blood—the corpuscles; these collect in the blood-vessels of the brain, and keep the blood from flowing as fast as it ought, which causes disease and very often death. Sometimes the brain is so much injured by the poison that the drinker becomes crazy, and is a great deal of trouble to himself and everybody else.

[91]

Since all this is true, wise children will let cider, lager, ale, wine, and every other kind of alcoholic drink alone, and never, NEVER,

> "Put an enemy into their mouths,
>
> To steal away their brains."

HARM DONE BY ALCOHOL TO THE

NERVES.	BRAIN.
Takes away their moisture, and paralyzes them.	Fills or congests its blood-vessels with impure blood.
Takes away their power to control the muscles.	Collects in it, and paralyzes it. Hardens its albumen.
Makes them unfit to carry messages to and from the brain.	So hurts it as to cause craziness (insanity) and death.

MORE ABOUT THE HARM DONE BY ALCOHOL.

In the lessons you have learned you have been taught about the harm done by alcohol to the body and the mind; can you tell, from what you have seen of drunken people, in what other way alcoholic liquors hurt them?—"They make people waste their money; they make them waste their time; they make them cross; they make them fight; they make them say silly and wicked words; they sometimes make fathers and mothers hurt their children; they make people lose their good name; they often make them do things for which they are sent to prison."

Yes, this is only some of the mischief done by alcohol. If you could fly around the world and see everybody who has been hurt in any way by this terrible poison, what a sad, sad sight you would behold! At least half the trouble in the world comes from strong drink.

Are *you*, little girl, little boy, going to join the army of drunkards? No, indeed! you think; but probably no one who has become a drunkard ever intended to do so. They all began [92] with one glass, a few drops of some alcoholic liquor,—cider, wine, or beer perhaps,—and thus learned to love the taste of alcohol, and soon became its slaves. For this poison has the strange power of making those who drink it want more and more of itself, though they know it is doing them harm.

The only safety is in letting alcoholic liquors alone, forever.

BLACKBOARD OUTLINE.

<u>ALCOHOLIC LIQUORS HURT</u>
The body,
The mind, and
The soul;
<u>AND MAKE PEOPLE</u>

WASTE	LOSE	UNFIT TO	UNFIT TO
Money,	Strength,	Think, or	SERVE
Talents, and	Health, and	Work.	Themselves,
Time.	Good name.		Their neighbor, or GOD.

STORIES ABOUT THE HARM DONE BY ALCOHOL.[6]

A YOUNG BEGINNER.—The hardest drinker I ever knew commenced on cider when he was only five years old. He would go to the barrel of cider in the cellar, which had been put there to make vinegar, and, getting a straw, would suck all the cider he wanted; and then, after he had played awhile, he would go back and get more. He kept on drinking alcoholic liquors of some kind, until he died a drunkard.

CIDER DELIRIUM.—Dr. J.H. Travis, of Masonville, N.Y., was once called to a child six years old, who was raving in the wildest delirium. His symptoms were so peculiar that he questioned the family closely, and found that the day previous, at a raising, the child had drank freely of cider. After the men left he had procured a straw and gone to the barrel and drank till he was senseless, and after this the delirium [93] came on. He exhibited undoubted symptoms of delirium tremens. Cider was the common beverage of the family. Dr. Travis has been called to several other cases of delirium tremens from the use of cider.—*Mrs. E.J. Richmond.*

A CAUTION TO MOTHERS.—One of the first literary men in the United States said to a temperance lecturer: "There is one thing which I wish you to do everywhere; entreat every mother never to give a drop of strong drink to a child. I have had to fight as for my life all my days to keep from dying a drunkard, because I was fed with spirits when a child. I thus acquired an appetite for it. My brother, poor fellow, died a drunkard."

A GIRL DRUNKARD.—A young girl of eighteen, beautiful, intelligent, and temperate, the pride of her home, was recommended to take a little gin for some chronic ailment. She took it; it soothed the pain; she kept on taking it; it created an artificial appetite, and in four years she died a drunkard.—*Medical Temperance Journal.*

"A LITTLE WON'T HURT HIM."—I was the pet of the family. Before I could well walk I was treated to the sweet from the bottom of my father's glass. My dear mother would gently chide with him, "Don't, John, it will do him harm." To this he would smilingly reply, "This little sup won't hurt him." When I became a school-boy I was ill at times, and my mother would pour for me a glass of wine from the decanter. At first I did not like it; but, as I was told that it would make me strong, I got to like it. When I became an apprentice, I reasoned thus: "My parents told me that these drinks are good, and I cannot get them except at the public-house." Step by step I fell.... I have grown to manhood, but my course of intemperance has added sin to sin. My days are now nearly ended. Hope for the future I have none.—*Dying Drunkard.*

DANGER.—In one of Mr. Moody's temperance prayer meetings at Chicago, a reformed man attributed a former relapse of drunkenness wholly to a physician's prescription to take whiskey three times a day!

KILLED BY THE POISON.—Many years ago, when stage coaches were in use in England, during a very cold night, a young woman mounted the coach. A respectable tradesman sitting there asked her what induced her to travel on such a night, when she replied that she was going to the bedside of her mother, of whose illness she had just heard. She was soon wrapped in such coats, etc., as the passengers could spare, and when they stopped the tradesman procured her some [94] brandy. She declined it at first, saying she had never drank spirits in her life. But he said, "Drink it down; it won't hurt you on such a bitter night." This was done repeatedly, until the poor girl fell fast asleep, and when they arrived in London she could not be roused. She was stiff and cold in death, and the doctor, on the coroner's inquest, said that she had been killed by the brandy. — *Mrs. Balfour.*

IN CASE OF SHIPWRECK. — In the winter of 1796 a vessel was wrecked on an island of the Massachusetts coast, and five persons on board determined to swim ashore. Four of them drank freely of spirits to keep up their strength, but the fifth would drink none. One was drowned, and all that drank spirits failed and stopped, and froze one after another, the man that drank none being the only one that reached the house at some distance from, the shore, and he lived many years after that.

IT EXHAUSTS STRENGTH. — Concerning one cold winter when there were very severe snow-storms in the Highlands of Scotland, James Hogg, the poet, says: "It was a received opinion all over the country that sundry lives were lost, and a great many more endangered, by the administration of ardent spirits to the sufferers *while in a state of exhaustion*. A little bread and sweet milk, or even bread and cold water, proved a much safer restorative in the fields. Some who took a glass of spirits that night never spoke another word, even though they were continuing to walk and converse when their friends joined them. One woman found her husband lying in a state of insensibility; she had only sweet milk and oatmeal cake to give him, but with these she succeeded in getting him home and saving him." — *Bacchus.*

SHIPMASTER OF THE KEDRON. — "I was brought up in a temperance school, and when I shipped before the mast I stuck to my principles, though everyone else on board drank excepting two boys whom I persuaded to abstain. In a very severe storm off a lee-shore, when it was so cold they had to break the icicles off the ropes to tack the ship, all drank but myself and these two boys. The men would work very well for a few minutes, and then slack off and take another drink, until they were all keeled up, and we three boys had all we could do to keep the ship from going ashore. If we had drank with the rest, all would have been lost, for the men were too drunk to save themselves. Providentially, the storm abated before morning, and we were saved. Now, for many years I have been captain of my own ship, and I never give out one drop of liquor." — *Captain Brown.*

[95]

ON THE PLAINS.—Twenty-six men, travelling on one of the great Western plains in the United States, were overtaken by cold and night. They had food, clothing, and whiskey, but no fire. They were warned not to drink whiskey or they would freeze. Three did not drink a drop, and though they felt cold they did not suffer nor freeze. Three more drank a little, and though they suffered much they did not freeze. Seven others that drank a good deal had their toes and fingers frozen. Six that drank pretty strong were badly frozen and never got over it. Four that got very boozy were frozen so badly that they died three or four weeks afterward. Three that got dead drunk were stiff dead by daylight. They all suffered just in proportion to the amount of whiskey they took. They were all strong men, and had about the same amount of clothing and blankets; the whiskey was all that made the difference.

THE RED RIVER EXPEDITION in Canada, in 1870, is often quoted as one of the most laborious on record, 1200 troops travelling 1200 miles through a very dense wilderness, and having all their supplies to carry. They were ninety-four days out, and none of them had liquor. They were constantly wet through, sometimes for days together, and all the while at the severe labor of rowing, poling, tracking, and portaging, yet they were always well and cheery, and there was a total absence of crime.

IN AFRICA it is far safer to do without intoxicating drink. Livingstone says that he lived without it for twenty years. Stanley performed his wonderful journey without it. Bruce said more than one hundred, years ago: "I laid down as a positive rule of health that spirits and all fermented liquors should be regarded as poisonous. Spring, or running water, if you can find it, is to be your only drink."

WATERTON, the great naturalist, who travelled so much in South America, says: "I eat moderately, and never drink wine, spirits, or any fermented liquors in any climate. This abstemiousness has proved a faithful friend." He died by accident at the age of eighty-three.

MR. HUBER, who saw 2160 perish of cholera in twenty-five days in one town in Russia, says that "Persons given to drinking are swept away like flies. In Tiflis, containing 20,000 inhabitants, every drunkard has fallen." Of 204 cases of cholera in the Park Hospital, New York, there were but six temperate persons, and these recovered. In Albany, where cholera prevailed with severe mortality for several weeks, only two of the 5000 members of temperance societies became its victims. [96] In Montreal, where the victims of the disease were intemperate, it usually cut them off. In Great Britain, those who have been addicted to spirituous liquors and irregular habits have been the greatest sufferers from cholera. In some towns the drunkards are all dead. — *Bacchus.*

MALT LIQUORS, under which title are included all kinds of porters and ales, produce the worst species of drunkenness. The effects of malt liquors are more stupefying than those of ardent spirits, and less easily removed. In a short time they render dull and sluggish the gayest disposition. — *Anatomy of Drunkenness.*

GINGER-BEER. — A man who has been a temperance-worker for forty-five years, says that there is often alcohol in ginger-beer. He told of a case known to him of a reformed man who, after drinking some, felt strongly drawn to the bar-room, where he drank until he brought on delirium tremens. The beer will sometimes ferment enough in a few hours to produce alcohol — if it answers the conditions — a sweet liquid and a ferment.

DANGER TO THE REFORMED. — A lady who had become a drunkard through taking alcoholic drinks as medicines, at length, after many efforts, succeeded in breaking away from the power of the appetite, and for a long time she seemed to be saved. At length she went to visit her mother, and that mother put brandy peaches on the table for tea. They aroused the slumbering appetite, the victim fell again, became worse than ever, and died a miserable drunkard.

[6] From *Juvenile Temperance Manual*, by Julia Colman.

STORIES ABOUT THE RIGHT WAY TO TREAT ALE, BEER, Etc.

THE RIGHT SIDE. — "Boys, which is the right side of the public house? Can you tell me?" — "Yes, sir, the outside."

THE GOAT AND THE ALE.—Many years ago, when everybody drank freely, a Welsh minister named Rees Pritchard was at the alehouse drinking, when he took it into his head to offer some ale to a large tame goat. The animal drank till he fell down drunk, and the minister drank on till he was carried home drunk. The next day he was sick all day, but on the third day he went again to the alehouse, and began to drink. The goat was there, and he offered him more ale, but the [97] animal would not touch it. The minister, seeing the animal wiser than himself, was ashamed, and gave up drinking, and became a worthy minister.

HOW THE MONKEY WAS CURED.—A monkey named Kees had been taught to drink brandy. At dinner every day he had his share like his more manly (?) neighbors, only that his was given to him in a plate. One day, as he was about to drink it, his master set it on fire, and he ran off frightened and chattering. No inducement could afterward make him drink brandy. We have many stories of animals who would never drink again after they had once experienced its effects.

THE KEEN MARKSMAN does not poison his nerves and brain with alcohol. Angus Cameron, a Highlander, at the age of twenty, took the Queen's prize for the best marksmanship, and when he was twenty-two (in 1869), he won in the same way a cup worth $1000. He made the best shot each time that ever had been made in the contest, and neither of them has been beaten by anyone else. Angus is a slight, modest, unassuming young man, who had been a Band of Hope boy. When he was announced as the winner, and all the friends made an ado over him, and offered him a generous glass of champagne, he quietly refused their mistaken kindness, and kept his pledge.

BENJAMIN FRANKLIN, when a printer boy in London, would drink no beer, and his companions called him the water American, and wondered that he was stronger than they who drank beer. His companion at the press drank six pints of beer every day, and had it to pay for. He was not only saved the expense, but he was stronger than they, and better off in every way. If he had gone to drinking beer at that time, like the other printer boys, it is likely we should never have heard of him.

OATMEAL DRINK.—"In Boulton and Watts' factory we saw an immense workman at the hottest and heaviest work, wielding a ponderous hammer, and asked him what liquor he drank. He replied by pointing to an immense vessel filled with water and oatmeal, to which the men went and drank as much as they liked." This is made by adding one pound fine oatmeal to each gallon of water, and is much used in factories and at heavy work of all kinds in Government works, instead of the old rations of alcoholic liquors. Iron puddlers, glass blowers, and athletic trainers, all do their work now better without alcoholic liquors.

[98]

A CHANGE IN AFFAIRS.—A poor boy was once put as an apprentice to a mechanic; and, as he was the youngest, he was obliged to go for beer for the older apprentices, though he never drank it. In vain they teased and taunted him to induce him to drink; he never touched it. Now there is a great change. Every one of those older apprentices became a drunkard, while this temperance boy has become a master, and has more than a hundred men in his employ. So much for total abstinence.

BOOKS BETTER THAN BEER.—An intelligent young mechanic stood up in a temperance meeting and said: "I have a rich treat every night among my books. I saved my beer money and spent it in books. They cost me, with my book-case, nearly $100. They furnish enjoyment for my winter evenings, and have enabled me, by God's blessing, to gain much useful knowledge, such as pots and pipes could never have given me."

A LITTLE DRUMMER-BOY was a favorite among the officers, who one day offered him a glass of strong drink. He refused it, saying that he was a Cadet of Temperance. They accused him of being afraid; but that did not move him. Then the major commanded him to drink, saying: "You know it is death to disobey orders." The little fellow stood up at his full height, and fixing his clear blue eyes on the face of the officer, he said: "When I entered the army I promised my mother on bended knees that, by the help of God, I would not taste a drop of rum, and I mean to keep my promise. I am sorry to disobey orders, sir, but I would rather suffer than disgrace my mother, and break my temperance pledge." He was excused from drinking.

[99]

TOBACCO.

INTRODUCTORY LESSON.

You have been learning about the poison alcohol, and what mischief is done by it; we will now study about another poison which thousands of persons are using every day. It is rolled in cigars and cigarettes, and hidden in snuff and pieces of tobacco, and does more harm to children and young people who use these things than to grown persons.

Perhaps you know how a person feels who takes tobacco or smokes a cigar for the first time; if not, we will tell you. He begins to be dizzy, to tremble, to become faint, and to vomit; his head aches, and he is so sick for hours, often for several days, that he scarcely knows what to do. Why is he so sick? Because tobacco poison has been taken into his lungs; also, some has mixed with the saliva and gone down into his stomach; and each part it has reached is striving to drive it out, and is saying, by the pain it causes, "You have given me poison; do not give me any more." If he had taken enough it would have killed him.

He recovers from this sickness and tries chewing or smoking again and again, until he becomes accustomed to the poison and can chew or smoke and it does not hurt him; so he thinks, but he is very much mistaken.

Tobacco is a poison, and hurts everybody who uses it every time they do so, although it does its evil work very slowly, unless taken in large quantities. To understand more about this we will try to learn how tobacco is obtained, what poison is in it, and in what way it harms people.

[100]

THE STORY ABOUT TOBACCO.

How it Came to be Used.—Tobacco is the leaves of the tobacco plant, a native of America. It was used by the Indians of this country before Columbus came here in 1492. Some of the Spaniards who were with him on his second visit took some of it back with them to Portugal, and told the people they had discovered a wonderful medicine. From Spain tobacco seed was sent to France by Jean Nicot, in 1560. It is said that Sir Walter Raleigh carried it to England in 1586, when Elizabeth was queen.

In a few years many civilized people were snuffing, chewing, and smoking tobacco, like the wild Indians, although it cost them a great deal of money to do so. King James does not seem to have liked it very much, for he said, "It is a custome loathsome to the eye, hateful to the nose, harmful to the brain, and dangerous to the lungs." He called the smoke "stinking fumes."

The Tobacco Plant. This plant belongs to the same family as the deadly nightshade, henbane, belladonna, thorn-apple, Jerusalem cherry, potato, tomato, egg-plant, cayenne pepper, bitter-sweet, and petunia. Most of the plants of this Nightshade family have more or less poison in their leaves or fruit. Tobacco is supposed to have been named from the pipe used by the Indians in smoking its leaves.

The common tobacco plant grows from three to six feet high, and has large, almost lance-shaped, leaves growing down the stems; its flowers are funnel-shaped and of a purplish color. When fresh the leaves have very little odor or taste.

How Tobacco is Used. — When the plants are ripe, they are cut off above the roots and placed where they will become dry, sometimes in a building made for this purpose, called "a tobacco house." After a short time they begin to smell strong and taste bitter. They are then stripped from the stems very carefully and sorted. The leaves nearest the root are considered the poorest, those at the top generally the best.

[101]

The different sorts are packed in separate hogsheads, and sent away to be sold to manufacturers of cigars, snuff, etc.

The manufacturer has some leaves rolled into cigars, some pressed into cakes for chewing, or into little pieces to be smoked in a pipe; while some are ground for snuff. While the dried leaves are being rolled, pressed, or ground, various substances are mixed with them to give them an agreeable odor and pleasant taste.

Yet, however pleasant the manufacturer may make them as he rolls, presses, or grinds, he cannot take the poison out of them. It remains in its brown covering to do much harm to those who may smoke the cigars, use the snuff, or chew the tobacco.

BLACKBOARD OUTLINE.

THE TOBACCO PLANT.

NATIVE OF America.	FOUND BY Columbus, 1492.	TAKEN TO Portugal, 1496. France, 1560.	GROWS IN THE Torrid and temperate zones.
(About 50 species.)		England, 1586.	

DESCRIPTION.

FAMILY

Height, 3 to 6 feet.
Leaves, lance-ovate, and running down the stem.
Stem, hairy and sticky.
Flowers, funnel-shaped and purplish.

The same as the Jerusalem Cherry,
Petunia,
Potato,
Tomato,
Egg-plant,
Red pepper, etc.

HOW MADE READY FOR USE.

(1)
Cut-off above the roots.
Dried.
Stripped; sorted.
Packed, and sold to the manufacturers.

(2)
Flavored and scented.
Rolled for cigars.
Pressed for chewing.
Ground for snuff.

[102]

THE POISON IN TOBACCO AND THE HARM IT DOES.

The Poison.—What is the poison in fermented liquors?—"Alcohol." In distilled liquors?—"Alcohol" True; and the strongest poison in tobacco is *nicotine,* named from the man who first sent it to France, Jean Nicot. Beside this it contains several others, some of which we shall tell you about when we make up our blackboard outline.

Tobacco, like alcohol, is a narcotic; that is, it soothes pain and produces sleep. Alcohol acts first upon the nerves; tobacco upon the muscles, which it weakens and causes to tremble. It often causes palpitation of the heart.

If the skin is scratched or punctured, and tobacco poison put into the wound, it will do the same harm as if it were taken into the stomach. Tobacco is so dangerous that physicians do not use it much as a medicine.

Harm done in the Stomach. — You remember that after alcohol has been swallowed, the little mouths of the stomach take it up and carry it to the liver, which sends it with the blood to different parts of the body.

Tobacco, as we have already told you, poisons more slowly. People do not swallow it purposely, yet some of it goes down, accidentally, into the stomach with the saliva, and makes trouble there, causing nausea and vomiting when taken for the first time. By and by the stomach seems to take the poison without being hurt, but it really suffers from dyspepsia or other diseases, and often loses its appetite for wholesome food.

Harm done in the Mouth, Throat, and Lungs. — The mouth takes in some of the poison through the pores of the membrane, or skin, which lines it; those who smoke, sometimes have what is called "smokers' sore throat"; besides this, the senses of taste and smell arc more or less injured by nicotine and the other poisons in tobacco.

The fumes, or smoke, from the weed fills the air with poisonous [103] vapor which irritates the lungs, not only of the smoker, but of all who are where they must breathe the same atmosphere. Lungs thus irritated are liable to become diseased.

Cigarettes are still more injurious than cigars because of the smoke from their paper coverings; also, because from the way they are made, more of the tobacco poison goes into the lungs. The cheap cigarette which boys use is made from cast-away cigar stumps and other filthy things.

Harm done in the Brain and Nerves. — The smoker feels so rested and comfortable, after his cigar, and his brain is so rested, that he does not think about the mischief that is going on among its blood-vessels and nerves; perhaps he has never heard that tobacco, snuffed, chewed, or smoked hurts the brain, and does not learn about it until he finds he is losing his memory, that his mind is not so strong to think as it should be, and his will too weak to help him conquer his love for the snuff, tobacco, or cigar, when he wishes to stop using it. He has become the slave of tobacco, and it is not easy to get free from his cruel enemy.

The nerves also lose their power, or become more or less paralyzed by nicotine and the other tobacco poisons.

More about the Harm done by Tobacco. — Some persons who continue to use tobacco are strong enough to throw off the poison through the lungs, the skin, and in other ways; but how much better it would be if they were not obliged to employ their strength in getting rid of that which does them no good, which only gives a little pleasure to nobody but themselves, and often makes those suffer who are compelled to remain where they are having "a good smoke." Beside, their breath and clothing have the tobacco odor, which not only makes the air impure, but is disagreeable to most people.

If this be true of smoking, what shall we say about the filthy habit of chewing, and the utterly useless and disgusting practice of taking snuff, which injures the voice as well as the senses of taste and smell? [104]

And what about spitting tobacco juice on the floors of cars, steamboats, churches, — any place where it is convenient for the man or boy who has lost his common politeness in his love for tobacco?

We must not forget that cigars, etc., cost money. No one who smokes, chews, or snuffs would throw away dollars and cents which might be put into the savings bank, or used in buying something worth having for himself or somebody else.

Lastly, we would have you know that tobacco causes thirst, and this often leads to drinking alcoholic liquors. Some one who has studied this subject, says that "nine out of ten of the boys and young men who become drunkards have first learned to smoke or chew tobacco." A New York daily paper gave a list of 294 cases of insanity caused by drinking, in 246 of which the whiskey drinking followed tobacco chewing.

Tobacco and alcohol make thousands of wretched homes, and send a great many people to prison or to the insane asylum; so we entreat you to turn from beer, wine, and all alcoholic liquors as you would from a serpent, and say No, when tempted to smoke a cigar or use tobacco in any form.

Do this all the more decidedly because, as we have told you before, alcohol and tobacco hurt children and young persons in every way more than they injure any one else. If you have begun to use these poisons, give them up this very day, before the habit of using them becomes too strong for you to break.

QUESTIONS ON THE USE OF TOBACCO.

Of what poison beside alcohol have you been studying?—"Tobacco."

How is tobacco used?—"Some take it in snuff; some chew it; some smoke it in a pipe; some smoke it in cigars or cigarettes."

What is the name of the strongest poison in tobacco?—"Nicotine."

What harm does tobacco poison do to the body?—See Blackboard Outline.

What harm does it do to the mind?—See Blackboard Outline.

Whom does it harm most?—"Those who begin to use it when they are children or very young."

What happens to children or young people if they use tobacco in any way?—"They are not healthy; they are not strong; they do not grow fast; they look pale and sickly."

[105]

How does the tobacco poison hurt their minds?—"They cannot learn fast; they often forget what they have learned."

What often makes tobacco-chewers, snuffers, and smokers disagreeable to clean people?—"Their breath smells of tobacco; their clothes smell of tobacco; they poison the air with tobacco-fumes; some have the filthy habit of spitting tobacco-juice wherever they happen to be."

What other harm does the use of tobacco do to people?—"It makes them waste time and money; it leads some to drink alcoholic liquors and to go with bad company."

If you are wise how will you treat tobacco?—"I will let it alone."

If you have begun to use it what had you better do?—"Give it up to-day."

Why to-day?—"Because the longer I use it the harder it will be for me to give it up."

If you keep on using it what will you be?—"A tobacco slave."

[106]

BLACKBOARD OUTLINE.

TOBACCO.

POISONS IN TOBACCO SMOKE.	EFFECTS OF THE POISONS.
Carbonic acid	Causes sleepiness and headache.
Carbonic oxide	Causes trembling of the muscles and heart.
Ammonia	Bites the tongue; makes too much work for the salivary glands.
Nicotine	See below.

NICOTINE

IS	CAUSES
Odorous,	Weakness,
Pungent,	Nervousness,

Emetic, Dizziness,
Poisonous, Nausea,
Pain-soothing, Faintness,
Sleep-producing, *i.e.* Narcotic. Loss of strength,
 Stupor,

If taken in large quantities Convulsions and Death.

SOME OF THE HARM DONE BY TOBACCO

TO THE BODY.	TO THE MIND, ETC.
Poisons the saliva.	Makes the memory poor.
Injures the sense of smell, taste, sight, and hearing.	Lessens the power to think.
	Weakens the will.
Causes "smokers' sore-throat."	Makes people grow in selfishness and impoliteness.
Injures the stomach, causing dyspepsia, etc.	Makes people waste time and money.
Often takes away the appetite for wholesome food.	Often leads to drunkenness and bad company.
Irritates the air-cells of the lungs.	Sometimes causes insanity.
Causes palpitation of the heart.	
Weakens the muscles, causing trembling.	
Injures the eyes.	
Excites, then stupefies and paralyzes the brain and the nerves.	

[107]

OPIUM AND OTHER NARCOTICS.

Opium. —Opium is the juice obtained from the seed-vessels of the white poppy before they are ripe; this is dried, and smoked in a pipe or chewed. It makes a person feel very pleasant and happy for a little while, then so horribly wretched that he takes more of the poison to forget his misery. So he keeps on until mind and body are a complete wreck. Now and then an opium slave gets free from the dreadful habit which has mastered him, but usually the slavery ends only in death.

Laudanum and Morphine. — These soothe pain and cause sleep; but beware of them; they are made from opium, and like it, though more slowly, hurt mind and body.

Beware also of *chloral hydrate* and *chloroform*, which physicians give to ease suffering and produce sleep. *Endure pain* rather than form the habit of using these narcotics.

Hashish, etc. — This is prepared from the hemp plant growing in hot countries, and is a terribly exciting poison.

The *areca nut*, the seed from a kind of palm, pear-shaped, and resembling a nutmeg, is mixed with quick-lime and wrapped in a betel-leaf, which grows on a vine belonging to the pepper family. This mixture reddens the saliva and lips, and blackens the teeth. It is chewed by millions of people in India.

The leaves of the *coca*, also of the *thorn apple*, are smoked or chewed by the South American Indian.

All these poisons mean the same thing, —

A little pleasure, DISEASE, and **DEATH**.

[108]

Practical Work in the School-Room.

BY SARAH F. BUCKELEW & MARGARET W. LEWIS.

Part I.—THE HUMAN BODY.

TEACHERS' EDITION.

A Transcript of Lessons given in the Primary Department of Grammar School No. 49, New York City.

This work was prepared especially to aid Teachers in giving oral instructions in Physiology to Primary and Intermediate Classes. It is, perhaps, the only Physiology published that is suitable for these grades. Considerable attention is paid to the subject of Alcohol and Narcotics.

> "First is given *a model lesson;* second, *a formula,* embodying the principal facts given during the development and teaching; third, *questions for the formula;* fourth, *directions for teaching;* and fifth, *questions on the lesson.* These last are important. A full plan of lessons is given for each week for five months, in each of six grades, showing exactly how much work ought to be attempted. No book could be made more helpful to teachers. To the thousands who are asking, 'Tell us how to teach,' here are full, minute, and correct instructions. Even the answers expected are given, blackboard outlines are arranged, and nothing is wanting to make the book as useful to teachers as it is possible for any book to be. It ought to have a large sale. No book published during the last ten years will do more to drive away routine from the school-room and introduce thought than this, *if only the teachers will use it.* Its introduction displaces nothing but the old-fashioned monotonous recitations. Let them go; we welcome this book as an important aid in hastening along the good time of better teaching. It is excellently printed, with good paper and binding."—*The New York School Journal.*

Illustrated. Price by mail, 75 cents.

DEVELOPMENT LESSONS.
BY PROF. E.V. DEGRAFF & MISS M.K. SMITH.
IN FIVE PARTS.

I. Fifty Lessons on the Senses, Size, Form, Place, Plants, and Insects.

> These lessons are presented objectively with a view to showing how elementary work in natural science may be done.

II. Quincy School Work.

III. Lectures on the Science and Art of Teaching.

> Specific instruction is given on how to teach Reading, Spelling, Phonics, Language, Geography, Arithmetic, etc.

IV. School Government.

V. "The New Departure in the Schools of Quincy." By CHAS. FRANCIS ADAMS.

> DR. A.D. MAYO says, in the *New England Journal of Education*: "Although we have given place in our book-notice column to an appreciative mention of the volume, 'Development Lessons,' a new reading seems to call for a new commendation of this admirable guide to teachers. Mr. DeGraff needs no special 'boom' as a first-class institute man, and his extracts of lectures in Part III. sparkle with valuable suggestions. In no published work is Col. Parker really seen to such advantage as in the 'reports of conversations' with him in Part II., which can be studied with profit by every teacher. But perhaps the most complete portion of this admirable book is the 178 pages of lessons on the Senses, Size, Form, Place, Plants, and Insects, by

MISS M.K. SMITH, now Teacher of Methods in the State Normal School at Peru, Neb."

Handsomely Bound and Illustrated. 300 pages. Price by mail, $1.50.

www.ingramcontent.com/pod-product-compliance
Lightning Source LLC
Chambersburg PA
CBHW031422210526
45464CB00005B/2004